THE COMMONWEALTH AND INTERNATIONAL LIBRARY
Joint Chairmen of the Honorary Editorial Advisory Board
SIR ROBERT ROBINSON, O.M., F.R.S., LONDON
DEAN ATHELSTAN SPILHAUS, MINNESOTA

SELECTED READINGS IN PHYSICS
General Editor: D. TER HAAR

NUCLEAR REACTIONS

Contents

Preface   vii

Part 1

CHAPTER I   EARLY SUCCESSES AND DIFFICULTIES   3
     1.1 Simple Quantitative Ideas   7
     1.2 Early Applications of Quantum Mechanics   10
     1.3 Reactions with Neutrons   16

CHAPTER II   THE COMPOUND NUCLEUS   20
     2.1 The Theory of Breit and Wigner   20
     2.2 Resonances in Nuclear Reactions   24
     2.3 Statistical Properties of Resonances   34
     2.4 The Statistical Model   37
     2.5 Mean Cross-sections: the Continuum Model   41
     2.6 Experiments and Their Analysis up to 1952   46

CHAPTER III   THE OPTICAL MODEL   53
     3.1 The Explanation of Feshbach, Porter and Weisskopf   53
     3.2 Elastic Scattering Experiments and Optical Model Fits   57
     3.3 Particle Propagation in the Optical Model   69

CHAPTER IV   NUCLEAR STRUCTURE AND NUCLEAR FORCES   77
     4.1 Nuclear Matter   78
     4.2 The Shell Model   83
     4.3 Collective Models   87

CHAPTER V	DIRECT INTERACTIONS	92
	5.1 Stripping	93
	5.2 Inelastic Scattering	100
	5.3 The $(p, 2p)$ Reaction	113

Bibliography   119

Part 2

1. "Discussion on the Structure of Atomic Nuclei", Sir Ernest Rutherford, O.M., F.R.S.   125
2. "The Theory of the Effect of Resonance Levels on Artificial Disintegration", N. F. Mott   136
3. "Neutron Capture and Nuclear Constitution", Niels Bohr   152
4. "Capture of Slow Neutrons", G. Breit and E. Wigner   165
5. "Fluctuations of Nuclear Reaction Widths", C. E. Porter and R. G. Thomas   192
6. "The Scattering of High-energy Neutrons by Nuclei", S. Fernbach, R. Serber and T. B. Taylor   216
7. "Regularities in the Total Cross-sections for Fast Neutrons", H. H. Barschall   224
8. "Model for Nuclear Reactions with Neutrons", H. Feshbach, C. E. Porter and V. F. Weisskopf   227
9. "Nuclear Reactions at High Energies", R. Serber   272
10. "Angular Distribution in (d, p) and (d, n) Reactions", A. B. Bhatia, K. Huang, R. Huby and H. C. Newns   276
11. "Elastic and Inelastic Diffraction Scattering", J. S. Blair   298
12. "Information Obtainable from $(p, 2p)$ Reactions", I. E. McCarthy   313

Index   325

Preface

FOR over fifty years nuclear reactions have been the primary source of a rapidly evolving mass of information about nuclei. The study of the mechanisms of the reactions themselves is a fascinating one, not only because of the information about nuclei that it yields, but because of the great diversity of quantum phenomena that must be understood and reconciled. In fact, until very recent experimental developments in atomic physics, nuclear physics was the only field in which energy measurements could be made with sufficient precision to study quantum scattering phenomena in terms of pure states. We have now reached the stage where there is hope that certain reactions can be understood largely in terms of the basic forces between nucleons.

In this book I have tried to explain the development of the understanding of nuclear reactions in a way that is intelligible to undergraduate students after a first course in quantum mechanics. The theory of nuclear reactions involves quite sophisticated applications of quantum mechanics, so that this is a very difficult task. However, I have tackled the problem head-on. Rather than attempt over-simplified explanations, I have attempted to develop the required knowledge and feeling for quantum mechanics in the course of explaining the most important phenomena. The task is simplified by the extremely interesting nature of the subject itself and of the original papers in Part 2, which are usually meant to be read concurrently with the text.

The choice of original papers has been motivated only partly by the requirements of space and simplicity. The key papers in the subject are long and require detailed study. The text has been written as an aide to this detailed study. The examples of the modern study of direct interactions are chosen with more regard to their power of explanation than to their place in the chronological order of the subject.

The present volume is intended as a companion to the earlier book in this series by Brink entitled *Nuclear Forces*, in the sense that I have referred to Brink's book whenever the question of the details of nuclear forces arises. However, it is self-contained if the most important properties of nuclear forces, which are mentioned here, are taken for granted.

I would like to thank my friends and colleagues whose help and encouragement have made the writing of this book possible. I would like to thank the publishers of the following journals for permission to reproduce the original articles in Part 2: *The Proceedings of the Royal Society, Nature, The Physical Review, The Philosophical Magazine, Reviews of Modern Physics*, also the University of Toronto Press. I am grateful to the authors of these articles and to Professor A. Bohr for confirming the permission.

The book was written concurrently with my research program sponsored by the Air Force of Scientific Research, Office of Aerospace Research, United States Air Force. I am grateful to this agency for its support.

PART 1

I

Early Successes and Difficulties

THE early history of nuclear physics may be divided into two periods of approximately twenty years. The first period began with the discovery of the nucleus by Rutherford at Manchester in 1911. It was characterized by years of hard, painstaking work and brilliant insight, inspired largely by Rutherford. Monoenergetic beams of α-particles from radioactive sources were used as probes for nuclear properties. Together with detailed information about the mass relationships between nuclei, these early reaction studies laid the foundation of knowledge about the nucleus. The second period began in 1932 with the discovery of the neutron and the invention of accelerators which enabled different probes to be used. During this period a basic understanding of nuclear reactions and structure in terms of quantum mechanics was achieved, but technology had not reached the stage where critical experiments to test the understanding could be performed or analysed. In the early 1950's more sophisticated accelerators, counting and data recording techniques and computing methods enabled the quantitative phase of nuclear physics to begin.

The earliest information about nuclei was obtained by studying the deflection of α-particles which occurs when they approach nuclei. When a particle hit a scintillating screen it caused a flash of light, which was observed by eye and manually recorded. The transfer of momentum to each α-particle was known by the direction of the emitted particle and its energy, which was first measured by using the fact that particles have short ranges in air. The ranges were calibrated against the energies from known sources. The subject of this book will be the deduction of nuclear properties from the distributions of momentum transfers to one or more particles in a nuclear reaction. The work of the first twenty years provides several interesting and important examples.

The other major source of information about nuclei was the measurement by Aston (1927) of nuclear masses with an accuracy of one ten-thousandth of the proton mass or 0·1 million electron volts (MeV) in energy. Study of the radioactive decay mechanism itself provided yet another source of information. The early developments in the study of nuclear reactions were summarized and explained by Rutherford (1929) in the opening address of a Royal Society discussion on the structure of atomic nuclei which is reproduced in Part 2. At this stage the known particles were the α-particle, the proton and the electron, although the existence of the neutron had been conjectured by Rutherford to explain the difference between the nuclear charge and mass numbers.

Successes with the α-particle probe came rapidly. In the earliest work the energies of the α-particles were not resolved and it was assumed that they were elastically scattered. The nucleus was discovered by the fact that the probability of finding an α-particle scattered at a certain angle with a certain energy is given by the Rutherford law for scattering by a heavy point nucleus where the force is the Coulomb repulsion. Knowledge of the mass of the nucleus confirmed this interpretation very accurately. The Rutherford law was obtained from classical mechanics. Larger scattering angles are due to particles approaching the nucleus more closely and scattering angles for the same distance of approach are smaller for higher energies.

Rutherford's group at the Cavendish Laboratory in Cambridge were able to observe scattering at backward angles, in practice about 135°, for which in the case of high enough energies and small enough nuclear charges the Coulomb scattering law broke down. It was assumed that the particle had come close enough to the nucleus to enter a region where the law of force was not the Coulomb law, but was given by some other strong, short-range effect. Measurement of the energy at which the Rutherford law broke down, which was not very well defined because of diffraction effects due to the quantum nature of the process, enabled an order of magnitude estimate of the nuclear radius to be obtained. This was the first use of a nuclear reaction to obtain information about nuclei.

Very soon it became possible to obtain deeper information about the energy relationships of nuclei by observing events in which an α-particle was absorbed by the target nucleus with the emission of a proton or a γ-ray or both. By measuring the energies of all the particles concerned and comparing them with Aston's mass measurements the relationship between mass and energy was confirmed. A reaction in which a particle a and target T interact to produce a final state particle b and residual nucleus R is called a $T(a, b)R$ reaction. We use the word "particle" to refer to the nucleus which is accelerated or detected.

One of the earliest reactions studied was Al^{27} $(α, p)$ Si^{30}. In addition to elastically scattered α-particles, two groups of protons, each characterized by a different energy, were observed. The energy of the first group corresponded to the mass difference between the initial and final particles and nuclei. The energy of the second group was lower, indicating that Si^{30} was left in an excited state which subsequently decayed to the ground state with the emission of a γ-ray of a certain energy. γ-rays of the right energy were in fact also found among the reaction products.

The use of the α-particle probe to discover excited states of nuclei developed very rapidly and led to the modern study of nuclear spectroscopy. Here interest in a reaction is limited to the fact that it will produce the required excited state and perhaps enable some of its properties to be determined. The spin, parity and decay rates of the states are examined by observing the radiations that they emit. We will be concerned with a deeper understanding of the mechanism of the reaction itself. Sometimes this understanding leads to new ways of obtaining spectroscopic information.

Preliminary information with great significance came again from the reaction Al^{27} $(α, p)Si^{30}$, this time concerning the probability of the α-particle reacting with the target. Chadwick, Constable and

Pollard (1931) found that the reaction occurred preferentially at four different energies between about 4 MeV and 5·3 MeV. Resonance was said to occur in the α,Al^{27} system at these energies. The widths of the resonances were about 0·25 MeV. Clearly the variation of reaction probability with energy and angle may be expected to yield information about nuclear structure.

The first radical change in our knowledge of nuclei occurred in 1932 with the identification by Chadwick of neutrons in the reaction $Be^9(\alpha,n)C^{12}$. Not only does the discovery of the neutron give us a starting point for the understanding of nuclear structure. It also provides us with a nuclear probe which has the valuable property that it reacts only with nuclear matter and not with the Coulomb field so that nuclear reactions at very low energy may be studied.

Another development occurred in 1932 which pioneered the study of reactions. This was the artificial acceleration of charged particles by Cockcroft and Walton (1932) at the Cavendish Laboratory. At first protons were accelerated to 0·6 and 0·8 MeV. The first artificially produced nuclear reaction was $Li^7(p,\alpha)He^4$. The invention of accelerating machines promised new probes, for example protons, deuterons, and even heavier ions, higher beam intensities and, more important, higher energies.

The Cockcroft–Walton machine accelerated particles in a single step from a terminal at ground potential to a high potential terminal for which the steady potential was provided by a bank of condensers and a half-wave rectifier. The single acceleration idea was developed independently at Princeton by Van de Graaff whose machine provided the high potential by means of an electrostatic charge delivered by a belt. Early Van de Graaff accelerators could accelerate protons to 1 MeV. A 2·5 MeV machine was built at Princeton in 1936.

A machine which promised much higher energies was the cyclotron. Protons are confined to closed orbits by a magnetic field and accelerated in one or two small steps for each turn by a radio-frequency potential from which they are shielded when it is in the wrong direction for acceleration. The cyclotron was invented in

1929 by Lawrence at the University of California at Berkeley. The first model produced a beam of 80 KeV protons in 1930. Lawrence and Livingstone (1932) built a larger machine which was ready in 1932 to accelerate protons to about 1 MeV. Their first studies of nuclear reactions were published in that year, shortly after those of Cockcroft and Walton.

In this chapter we will discuss quantitative ideas which were developed to explain reactions before 1932 and we will see the effect of the development of the low-energy neutron probe.

1.1 Simple Quantitative Ideas

Given a monoenergetic collimated beam one knows the initial momentum in a reaction. It is therefore of interest to know the energy and direction of the emergent particle (or particles) and of course the probability of finding a particle emerging with a certain momentum. We will mainly be concerned with reactions involving a single emergent particle.

The object of an experiment is first to identify the final state energy so that separate quantum states may be studied. A quantum state of the system consisting of the residual nucleus plus the emergent particle is called a channel, specifically an exit channel. The quantum state of the system consisting of the incident particle and the target is the entrance channel. For a particle with only one internal quantum state there is one channel for every quantum state of the nucleus. The α-particle has only one quantum state for low energies. Its first excited state is at about 20 MeV. Both the proton and neutron have spin $\frac{1}{2}$ and therefore two states of spin projection. They may be identified separately by measuring the spin polarization of the beams. This is a refinement which will be mentioned later. The spin-dependent forces are quite small and it is a good first approximation to treat protons and neutrons as if they have only one quantum state. A particular reaction is characterized by the entrance and exit channels.

If one knows the quantum states of the initial and final systems, one can find the probability of the reaction occurring as a function of the momentum transfer from the incident to the emergent particle. This probability is conveniently expressed in terms of the differential cross-section which is a quantity dependent only on the properties of the reacting particles and not on experimental conditions such as the intensity of the incident beam. The differential cross-section for a reaction at a given energy E is denoted $d\sigma(\theta, \phi, E)/d\Omega$ and defined by

$$dN = I\,[d\sigma(\theta, \phi, E)/d\Omega]\,d\Omega, \qquad (1.1)$$

where dN particles are scattered per second into an element of solid angle $d\Omega$ making an angle (θ, ϕ) with the incident beam. The incident beam intensity is I particles per unit area per second. In many cases the scattering centre is spherically symmetric so that the problem has axial symmetry. The angle ϕ need not then be specified.

The first example of nuclear information being obtained from measurements of the differential cross-section as a function of momentum transfer is the Rutherford law for elastic scattering, in which the system remains in the entrance channel. In this case

$$\frac{d\sigma(\theta)}{d\Omega} = \frac{1}{4}\left(\frac{ZZ'e^2}{2E}\right)^2 \operatorname{cosec}^4 \frac{\theta}{2}, \qquad (1.2)$$

where E is the incident energy in the centre of mass system, Z and Z' are the charge numbers of the target and probe respectively and e is the charge of the proton. Since E depends on the target mass M it is possible to obtain M from the experiment and verify either the Rutherford formula or an independent mass measurement.

The momentum transfer P in this experiment is given for the centre of mass system by

$$P = 2(2\mu E)^{1/2} \sin \theta/2, \qquad (1.3)$$

where $\mu = mM/(m+M)$ is the reduced mass for an incident particle of mass m. The differential cross-section as a function of momentum transfer is given by

$$\frac{d\sigma(P)}{d\Omega} = 4\mu^2(ZZ'e^2)^2 \, P^{-4}. \tag{1.4}$$

The unit of length in nuclear physics is the fermi (fm) which is 10^{-13} cm. Differential cross-sections, however, are customarily expressed in millibarns (mb) per steradian (sr) where 1 mb = 10^{-1} fm^2 = 10^{-27} cm^2. Cross-sections integrated over angles are expressed in barns. 1 barn = 10^{-24} cm. The unit of energy is 1 MeV.

If the Rutherford law breaks down for scattering angles greater than θ_0 we say that the radius of the nucleus is equal to the impact parameter of the trajectory whose asymptote makes an angle θ_0 with the incident direction.

$$R = \frac{ZZ'e^2}{2E} \cot \frac{\theta_0}{2}. \tag{1.5}$$

We see that for a given scattering angle particles of higher energy come closer to the nucleus. An idea of the magnitudes involved is obtained by calculating the radius of Al^{27} for an experiment in which anomalous scattering at 135° sets in at 0·9 MeV. The original experimental curve for this reaction is shown in Rutherford's address in Part 2. The value $R = 6$ fm is obtained very quickly from (1.5) by making use of two well-known quantities, the mass of the electron

$$m_e c^2 = 0 \cdot 511 \text{ MeV}$$

and the classical radius of the electron

$$e^2/m_e c^2 = 2 \cdot 82 \text{ fm}.$$

There is considerable uncertainty in the determination of the critical energy. Modern experiments with high-energy electron probes are analysed by quantum theory to yield smaller values of R.

In order to observe nuclear properties it is necessary that the probe be able to enter the nucleus. This means that it must have at least as much energy as the value of the Coulomb potential at

R. This value is called the Coulomb barrier. For values of R found by Rutherford (about $2A^{1/3}$ fm) the height of the Coulomb barrier for α-particles is given by

$$V_\alpha \simeq 1\cdot 14 Z^{2/3}.$$

Here we have assumed that $A = 2Z$. For protons the Coulomb barrier is half this value

$$V_p \simeq 0\cdot 57 Z^{2/3}.$$

Although wave mechanical barrier penetration effects mean that the Coulomb barrier is not an absolute lower limit on the useful energy of a probe, the cross-section for a nuclear reaction falls very rapidly as the energy decreases below this value. Thus reactions initiated by 6 MeV α-particles or 3 MeV protons are usefully studied only for targets lighter than about Al. This in itself provides strong motivation for the development of accelerators of higher energy.

1.2 Early Applications of Quantum Mechanics

The interaction of the probe with the target becomes interesting when the probe can penetrate the Coulomb barrier. The nuclear forces then act on it. The simplest assumption about the effect of the nuclear forces is that they may be represented by a spherical one-body potential $V(r)$ which is different from the Coulomb potential inside the radius R. Since nuclei are bound we expect the potential to be attractive.

Because the quantum states of light nuclei are well-separated in energy and therefore easily identified, nuclear reactions were ideally suited to study by means of the newly discovered quantum mechanics.

The potential model could be tested either by using it to explain the anomalous backward scattering for a given energy, which may be several orders of magnitude above the Rutherford value for certain nuclei, or to explain the variation of cross-section with incident energy, which was first described as resonance behaviour by Gurney (1929). In fact the latter approach proved more interesting.

It was soon shown by several authors that simple one-dimensional wave mechanical ideas predicted resonances whose energy and width were of the right order of magnitude, about 1 MeV. Resonances in one-dimensional potential wells are due to standing waves caused by reflection of the incident wave from the regions of changing potential. At energies for which the wave function is large internally the probability of penetration, and hence of a reaction occurring, is high. Resonances are analogous to stationary states for which there are certain energies where the internal wave function is non-vanishing.

A calculation in three dimensions was performed by Mott in 1931. His paper is reproduced in Part 2 of this book. The first sections are valuable as an introduction to the method of solving the Schrödinger equation for the scattering of a spinless particle by a potential in the presence of the Coulomb force. We will outline the main results of this method in modern notation since it is fundamental in nuclear reaction theory.

The scattering depends on the Coulomb parameter α (which is called η in much of the later literature).

$$\alpha \equiv \eta = ZZ'e^2/\hbar v, \quad (1.6)$$

where $\hbar = h/2\pi$ and v is the velocity of the incident particle in the centre of mass system.

The Schrödinger equation for the scattering is

$$\nabla^2 \psi(r) + \frac{2\mu}{\hbar^2}[E - V(r)]\psi(r) = 0. \quad (1.7)$$

It may be separated in radial and angular coordinates yielding the general solution

$$\psi(r) = 4\pi(kr)^{-1} \Sigma_{lm} i^l e^{i\sigma_l} u_l(k,r) Y_l^m(\Omega_r) Y_l^{m*}(\Omega_k). \quad (1.8)$$

Ω_r and Ω_k are the directions of r and the incident wave vector k, where the de Broglie wave number is $k = [2mE/\hbar^2]^{\frac{1}{2}}$. Alternatively, choosing the z axis for spherical polar coordinates to be in the direction of the unit vector $\hat{\mathbf{k}}$ which is the incident direction,

$$\psi(r) = (kr)^{-1} \Sigma_{lm} i^l (2l + 1) e^{i\sigma_l} u_l(k, r) P_l(\cos \theta). \qquad (1.9)$$

$Y_l^m(\Omega)$ is a spherical harmonic and $P_l(\cos\theta)$ is a Legendre polynomial. The radial wave function $u_l(k, r)$ is the solution of

$$\frac{d^2 u_l}{dr^2} + \left\{\frac{2\mu}{\hbar^2}[E - V(r)] - \frac{l(l+1)}{r^2}\right\} u_l = 0. \qquad (1.10)$$

The phase factor σ_l is the Coulomb phase shift (see equations (1.15, 1.16)). The function u_l differs from Mott's function L_l by the fact that kr has been multiplied in to simplify the radial equation (1.10) in comparison with Mott's equation (4).

If $V(r)$ does not include a Coulomb potential, the asymptotic form of the wave function for large r is given by Huygens' principle. The plane wave interferes with a spherical outgoing wave whose amplitude is $f(\theta)$.

$$\begin{aligned}\psi(r) &\sim I + f(\theta)S/r \\ &= e^{i\mathbf{k}\cdot\mathbf{r}} + f(\theta)\, e^{ikr}/r.\end{aligned} \qquad (1.11)$$

The amplitude $f(\theta)$ is the scattering amplitude. Its square is easily seen to be equal to the differential cross-section.

When the long-range Coulomb potential is included it distorts the waves at infinity. For a pure Coulomb potential

$$I = \exp\left[i\mathbf{k} \cdot \mathbf{r} + i\alpha \log kr\,(1 - \cos \theta)\right], \qquad (1.12)$$

$$S = \exp\left[ikr - i\alpha \log 2kr\right], \qquad (1.13)$$

and the scattering amplitude is

$$f^c(\theta) = \frac{ZZ'e^2}{2\mu v^2}\operatorname{cosec}^2 \frac{\theta}{2} \exp\left(-i\alpha \log \sin^2 \frac{\theta}{2} + 2i\sigma_0 + i\pi\right). \qquad (1.14)$$

The square of this amplitude is the Rutherford cross-section (1.2). In the case of the Coulomb potential, classical and quantum mechanics give the same cross-section.

The expansion (1.9) is called a partial wave expansion. The separation constants l and m give rise to spherical harmonics $Y_l^m(\theta, \phi)$ which are eigenstates of the orbital angular momentum operators l^2 and l_z, just as for the hydrogen atom problem. Thus each partial wave corresponds to an orbit (or trajectory) whose angular momentum is l in the equivalent of the Bohr model of the atom for unbound states. For spherical problems without spin interactions l is a good quantum number. However, in the present axially symmetric problem the partial wave expansion is just a convenient mathematical device. It is instructive to notice that large values of l correspond to trajectories which miss the nucleus. Therefore the effect of the nuclear forces is noticeable only in the first few partial waves up to about $l_0 = kR$. A very useful number to remember is the energy of a proton for which $k = 1$ fm^{-1}. This is 20·75 MeV. For other particles this can be scaled according to the square root of the mass. Using this number we can easily determine which partial waves are relevant to a scattering calculation.

For Coulomb scattering the radial wave function $u_l(k, r)$ is the Coulomb function $F_l(\alpha, k, r)$ which is regular at the origin. Its asymptotic form is

$$F_l(\alpha, k, r) \sim \sin(kr - l\pi/2 - \alpha \log 2kr + \sigma_l). \qquad (1.15)$$

Since (1.10) is a second order equation there is a second, irregular, solution $G_l(\alpha, k, r)$ given asymptotically by

$$G_l(\alpha, k, r) \sim \cos(kr - l\pi/2 - \alpha \log 2kr + \sigma_l). \qquad (1.16)$$

It is clear from (1.15, 1.16) why σ_l is called the Coulomb phase shift.

In solving a scattering problem one integrates (1.10) numerically up to a certain channel radius r_0 beyond which the potential has the pure Coulomb form. In the exterior region for $r > r_0$ the solution of (1.10) is a linear combination of F_l and G_l which is matched smoothly to the interior solution at r_0.

$$u_l(k, r_0) = F_l(a, k, r_0) + C_l[G_l(a, k, r_0) + iF_l(a, k, r_o)]. \quad (1.17)$$

The combination $G_l + iF_l$ is an outgoing spherical wave asymptotically, as is most easily seen in the case $a = 0$ from the forms (1.15) and (1.16). The matching coefficient C_l may be written

$$C_l = (e^{2i\delta_l} - 1)/2i, \quad (1.18)$$

where δ_l is called the nuclear phase shift since it appears in the asymptotic form of u_l as follows:

$$u_l(k, r) \sim \sin(kr - l\pi/2 - a \log 2kr + \sigma_l + \delta_l). \quad (1.19)$$

This may be verified after some algebraic manipulation involving (1.15, 1.16, 1.17, 1.18). The nuclear phase shift is a property of the nuclear potential and the incident energy. In Mott's calculation he first assumed that resonance occurred only in the s-state and that to a good approximation $\delta_0 \equiv P$ was the only non-vanishing nuclear phase shift.

The scattering amplitude in general is

$$f(\theta) = f^c(\theta) + \frac{1}{2ik} \Sigma_l (2l + 1)(e^{2i\delta_l} - 1)P_l(\cos\theta), \quad (1.20)$$

where the sum need only be taken as far as l_0 since $\delta_l = 0$ for $l > l_0$. Except at very low energies l_0 may be taken as about $kR + 3$. The generalization (1.20) may easily be made from the considerations of Mott's paper. A complete discussion of simple scattering theory may be found, for example, in Blatt and Weisskopf (1952).

A concept which is useful for scattering of uncharged particles is the total scattering cross-section obtained by integrating $|f(\theta)|^2$ over angles. This cross-section is infinite for charged particles. We use the relation

$$\int_{-\pi}^{\pi} \sin\theta \, d\theta \, P_l(\cos\theta)P_{l'}(\cos\theta) = [4\pi/(2l + 1)]\,\delta_{ll'} \quad (1.21)$$

to obtain

$$\sigma = \Sigma_l' \sigma_l,$$
$$\sigma_l = \pi \lambdabar^2 (2l+1) \, |\, 1 - \eta_l \,|^2, \quad (1.22)$$
$$\eta_l = e^{2i\delta_l}.$$

The de Broglie wavelength is
$$\lambdabar = \lambda/2\pi = 1/k. \quad (1.23)$$

In Mott's example the scattering amplitude is

$$f(\theta) = f^c(\theta) + \frac{1}{2ik}[e^{2iP} - 1]. \quad (1.24)$$

Mott gave the expected shape of the curve of the variation of differential cross-section at 180° with energy as the incident energy is varied over the resonance (Mott's figure 1). This shape is now well known for resonances in elastic scattering.

In section 3 of Mott's paper he considered the absorption of the α-particle by the nucleus with the conservation of momentum being satisfied by the subsequent emergence of a proton. This treatment represents the entrance channel and the reaction channel by wave functions whose forms are distorted from the free particle forms by one-body potentials representing the particle–nucleus interactions. It is called a distorted wave treatment. By considering the asymptotic form of the wave functions and the conservation of probability flux, Mott showed that there is an upper limit on the capture cross-section $a(E)$ for a resonance characterized by orbital angular momentum n.

$$a(E) \leqslant (2n+1)\pi \lambdabar^2. \quad (1.25)$$

Resonances in the absorption cross-section appeared to be weaker than in the scattering cross-section.

Mott finally worked out an example using plausible values of the physical quantities involved which gave a resonance width of a few tenths of a MeV in qualitative agreement with experiment, although experimental resonances were not fitted.

A resonance in elastic scattering at E_0 is understood as a quantum state of the system consisting of target plus incident particle at an energy E_0 which has uncertainty Γ because the system can decay. In an ensemble of systems the number dN which decay in time dt is proportional only to dt and the number N of remaining systems.

$$dN = -\tau^{-1} N\, dt. \tag{1.26}$$

If the number of systems in the resonant state at $t = 0$ is N_0, the relative number $N(t)/N_0$ after time t is given by solving (1.26).

$$N(t)/N_0 = \exp(-t/\tau). \tag{1.27}$$

The relative number may be called the decay probability of a single system. The time constant τ is the lifetime. It may be thought of as the time uncertainty in the existence of the resonant state. The corresponding uncertainty in energy $\hbar/\tau$ is called Γ, the width of the resonance.

The time taken by an α-particle moving at one-tenth of the speed of light to cross a nucleus of radius 6 fm is approximately 10^{-22} sec. For potential scattering this may be considered as the lifetime of the α-target system. The corresponding resonance width is

$$\Gamma = \hbar/(10^{-22}\text{ sec}) \simeq 1 \text{ MeV},$$

as was found by Mott in his more detailed calculation.

1.3 Reactions with Neutrons

By 1932 the outlook for an understanding of reactions appeared quite bright. The energy relationships were completely understood for ground states. The potential scattering model gave promise of being capable of refinement to predict cross-sections at different energies once sufficiently accurate data were available. The prediction of the nuclear excited states themselves from basic nucleon–nucleon forces would require considerable refinement in the methods of tackling many-body problems, and knowledge of the nucleon–nucleon force would have to be refined before this programme could be attempted. However, no serious anomalies had developed which constituted a barrier to understanding.

The development of the neutron probe after 1932 offered further opportunities for understanding reactions. A neutron beam was made by bombarding Be^9 with α-particles from radioactive sources, for example polonium. Unresolved neutron energies up to 11 MeV could be obtained in this way. Semiquantitative experiments with fast neutrons showed that all the reactions expected on the basis of energy conservation, for example (n, n), (n, p) $(n, α)$, $(n, γ)$ occurred with cross-sections comparable with or less than the geometrical cross-section of about 1 barn. The cross-section for $(n, γ)$, where the reaction occurred, was several times less than the elastic (n, n) cross-section, but not orders of magnitude below it, just as would be expected from a distorted wave theory such as that of Mott.

One advantage of the neutron probe is the fact that it will react with the nucleus at very low energies. Experiments with neutrons slowed down to thermal velocities by collisions with hydrogen atoms in water or paraffin wax were conducted by several groups. The two-channel distorted wave theory predicts that there are no strongly energy-dependent dynamic effects since the dynamics is described merely by collective one-body potentials for the interaction of the particles involved in the reaction with all the other particles in the nucleus. The variation of scattering and absorption cross-sections with mass number A is slow since effects such as resonances depend only on the size of the nucleus. The absorption probability per unit time is closely proportional to the density of incident neutrons outside the nucleus. The absorption cross-section is therefore proportional to $1/v$ where v is the relative velocity of the incident neutron and the nucleus.

Although the incident neutron energies were not resolved it was possible to obtain a rough check of the $1/v$ law by measuring average cross-sections. Such experiments were performed by Bjerge and Westcott (1935), Moon and Tillman (1936) and Amaldi *et al.* (1936). The results were totally unexpected. For some nuclei the capture cross-section was small or negligible, for others of neighbouring mass number it might be of the order of thousands of barns while the scattering cross-section was many orders of magnitude less.

These effects were immediately investigated in more detail with the object of finding the order of magnitude of the widths and spacings of the neutron resonances. Szilard (1935) showed that the resonances were far too narrow for a potential scattering model and that the low-energy absorption cross-section departed violently from the $1/v$ law. Frisch and Placzek (1936) presented evidence that for the lowest energies widths were of the order of eV and spacings were of the order of 10 eV.

It is immediately concluded that the lifetime of the n-target system is a million times longer than would be expected for potential scattering. Bohr, in a famous address to the Copenhagen Academy on January 27, 1936, gave a mechanism which qualitatively explains this, while at the same time introducing a note of extreme pessimism about the prospect of ever obtaining a quantitative mechanistic understanding of reactions. The many-body nature of the nucleus is responsible for the long-lived state. The neutron first collides with a particle in the nucleus and shares its energy so that no one particle has enough energy to leave the potential generated by the collective effect of all the protons and neutrons (collectively called nucleons). However, the compound nucleus thus formed has a total energy higher than the ground state so it must eventually decay either by re-emitting the neutron or by emitting a γ-ray. If a neutron is to be emitted we must wait for millions of collisions, which are of course quite unpredictable, until a statistical fluctuation comes along and one of the neutrons acquires enough energy to escape. The emission of a γ-ray, which is a collective effect of the proton charges, is much more probable in certain heavy nuclei. In nuclei with $A < 70$, radiation is very improbable.

Bohr conjectured that at higher energies, even up to hundreds of MeV, essentially the same considerations apply. A statistical fluctuation may occur in which a nucleon or a group of nucleons

EARLY SUCCESSES AND DIFFICULTIES

or several groups of nucleons in sequence have enough energy to escape. Bohr's statement of this philosophy is as follows. "In fact the essential feature of nuclear reactions, whether incited by collision or by radiation, may be said to be a free competition between all the different possible processes of liberation of material particles and of radiative transitions, which can take place from the semistable intermediate state of the compound system." This way of considering reactions has come to be called the Compound Nucleus Model or Strong Coupling Model.

II

The Compound Nucleus

LESS than three weeks after Bohr's address to the Copenhagen Academy, Breit and Wigner submitted a paper to *The Physical Review* which gave the quantum mechanical explanation of the scattering, capture anomaly and which contained the essentials of a complete theory of the compound nucleus. The theory depends on the parameters of resonances, an *a priori* calculation of which would require the complete solution of the many-body problem. The sense in which we use the term "Compound Nucleus" is wider than that of the compound nucleus model. The compound nucleus is the system consisting of the probe and the target or, equivalently, the final state particles, while they are sufficiently close to be affected by nuclear forces. No model is involved.

2.1 The Theory of Breit and Wigner

The paper of Breit and Wigner is reproduced in Part 2. We will give an expanded explanation of the starting point of the theory. For mathematical convenience the system is assumed to be enclosed in a large cubical box of volume V. The wave functions ϕ_s for the neutron, target channel and ϕ_r for the γ-ray, residual nucleus channel obey periodic boundary conditions at the walls of the box and thus have discrete eigenvalues.

The time-dependent wave function $\psi(t)$ of the whole system is expanded in terms of the complete set of all the wave functions of the system. These include the wave functions of all the scattering channels labelled by s and all the radiative channels labelled by r, in which the neutron and the γ-ray are well separated from the nucleus and do not interact with it.

The essential difference between the Breit–Wigner theory and the older two-channel distorted wave treatment is the inclusion in the theory of the wave functions ϕ_c of the quantum states of the compound system in which the neutron is in the nuclear region and the γ-ray has not been emitted. These are discrete quasi-stationary or resonant states. In this first paper Breit and Wigner made the approximation that only one of the compound states was at a sufficiently low energy to have a serious effect on the reaction. Strictly the complete set of wave functions includes those of all possible exit channels including different outgoing particles. At low energies these channels cannot be excited. They are said to be closed, and were also omitted by Breit and Wigner.

With these approximations the time-dependent wave function of the system is expanded as follows:

$$\psi(t) = \Sigma_s a_s(t)\phi_s + c(t)\phi_c + \Sigma_r b_r(t)\phi_r \tag{2.1}$$

The Hamiltonian is

$$H = (H_T + H_n + V_{nT}) + H_\gamma + V_\gamma$$
$$\equiv H_{nT} + H_\gamma + V_\gamma, \tag{2.2}$$

where the symbols on the right-hand side represent the Hamiltonians of the target and the neutron, the interaction between the neutron and the target, the radiative Hamiltonian and the radiative interaction. H_T, V_{nT} and V_γ are all many-body quantities. The Schrödinger equation representing the time development of the system is

$$H\psi(t) = i\hbar \frac{\partial}{\partial t} \psi(t). \tag{2.3}$$

Substituting the expansion (2.1) in (2.3) we have

$$\Sigma_s a_s H\phi_s + cH\phi_c + \Sigma_r b_r H\phi_r$$
$$= (ih/2\pi)(\Sigma_s \dot{a}_s \phi_s + \dot{c}\phi_c + \Sigma_r \dot{b}_r \phi_r). \tag{2.4}$$

The form of some of the matrix elements of H follows from the definitions of H and the state vectors. We will adopt an abbreviated notation for the numerical quantities as follows:

$$-\langle \phi_s | H | \phi_c \rangle = M_s = hA_s$$
$$-\langle \phi_c | H | \phi_r \rangle = M_r = hB_r$$
$$\langle \phi_{s'} | H | \phi_s \rangle = h\nu_s \delta_{ss'} \quad (2.5)$$
$$\langle \phi_{s'} | H | \phi_r \rangle = 0$$
$$\langle \phi_c | H | \phi_c \rangle = h\nu$$
$$\langle \phi_{r'} | H | \phi_r \rangle = h\nu_r \delta_{rr'}.$$

Three coupled differential equations for the amplitudes a_s, c and b_r are obtained by multiplying (2.4) on the left in turn by the Hermitian conjugates of ϕ_s, ϕ_c, and ϕ_r and integrating over the nuclear coordinates.

$$\left[\frac{1}{2\pi i}\frac{d}{dt} + \nu_s\right] a_s = A_s c$$
$$\left[\frac{1}{2\pi i}\frac{d}{dt} + \nu\right] c = \Sigma_s A_s^* a_s + \Sigma_r B_r b_r \quad (2.6)$$
$$\left[\frac{1}{2\pi i}\frac{d}{dt} + \nu_r\right] b_r = B_r^* c.$$

These equations are identical with equations (3) of Breit and Wigner.

Although equations (2.6) are not a complete set of coupled equations because the equations for scattering and radiative channels other than those for particular values of s and r have not been written, Breit and Wigner solve them approximately. Setting the mean spacings in frequency of the channels s and r as $\Delta\nu_s$ and $\Delta\nu_r$, the quantities Γ_s, Γ_r and Γ are defined by

$$\Gamma_s = [\pi \overline{|A_s|^2}/\Delta\nu_s]_{\nu_s = \nu_0}$$
$$\Gamma_r = [\pi \overline{|B_r|^2}/\Delta\nu_r]_{\nu_r = \nu_0} \quad (2.7)$$
$$\Gamma = \Gamma_s + \Gamma_r,$$

where ν_0 is the resonant frequency. In time-dependent perturbation theory Γ_s and Γ_r, so defined, are proportional to the scattering and reaction probabilities respectively. From the discussion of resonances in section 1.2 we remember that they are therefore proportional to the widths of resonances in those channels. The averages $\overline{|A_s|^2}$, $\overline{|B_r|^2}$ are directional averages, explained by Breit and Wigner in the discussion after their equation (12). If the state c is an s-state we do not need to take an average. This case will provide a sufficiently general example. For states of higher angular momentum the averaging operation is equivalent to multiplying $|A_s|^2$ or $|B_r|^2$ by a statistical factor S which determines the effectiveness of the coupling of the entrance channel to the state c. The initial conditions used for the solution at $t = 0$ are

$$a_s = \delta_{ss_0}, b_r = 0, c = 0, (t = 0). \tag{2.8}$$

The capture and scattering cross-sections σ_c and σ_s are obtained from studying the time development of the solutions of (2.6) after the initial transients have died down. The box is expanded to infinity at the end of the calculation. We have

$$\sigma_c = \frac{\lambda^2}{\pi} S \frac{\Gamma_s \Gamma_r}{(\nu - \nu_0)^2 + \Gamma^2},$$

$$\sigma_s = \frac{\lambda^2}{\pi} S \frac{\Gamma_s^2}{(\nu - \nu_0)^2 + \Gamma^2}. \tag{2.9}$$

The quantities Γ_s and Γ_r are called the partial widths (in frequency) for scattering and radiation. The total width Γ is the sum of the partial widths for open channels. In modern notation the quantities $2\hbar\Gamma_s$ and $2\hbar\Gamma_r$, which have dimensions of energy, are called partial widths. If the compound state c has an orbital angular momentum quantum number L, then it is explained by Breit and Wigner that the statistical factor S is $2L + 1$ in the absence of spin–orbit coupling.

Breit and Wigner point out that the expression for σ_s in (2.9) ignores the spherical scattered wave in the entrance channel s_0 which is often called the potential scattering term. This expression is therefore a good approximation when the resonance is very strong.

Equations (2.9) show how it is possible for Bohr's qualitative description of a low energy reaction to be realized in quantum mechanics. All nuclear states except the ground state can decay by γ-emission, although for some states the radiation width Γ_r is vanishingly small. The neutron decay width Γ_s can be extremely small since it may take a very long time for a neutron to acquire enough energy through a statistical fluctuation to escape from the nucleus. The scattering cross-section is many orders of magnitude less than the capture cross-section for those heavy nuclei for which Γ_r is large.

2.2 Resonances in Nuclear Reactions

If all possible compound states and exit channels are included in the expansion (2.1) it is clearly possible to write down an infinite set of coupled differential equations which constitute a complete formal theory of a reaction. However, this method does not lend itself to the development of a useful theory. Bethe (1937) derived the Breit–Wigner theory from a time-independent formalism, assuming well-separated resonances. Exact formal theories are time-independent. They have been developed by Kapur and Peierls (1938), Wigner and Eisenbud (1947) (the R-matrix theory), Bloch (1957), Humblet and Rosenfeld (1961) and others. The theories differ by selecting different representations for the compound states ϕ_c. Resonance parameters are defined in terms of the representations.

We will give the basic ideas of resonance theory according to the time-independent treatment outlined by Blatt and Weisskopf (1952). It will often be a sufficient illustration to consider reactions initiated by s-wave neutrons. We assume that a limited number of particle-decay channels, denoted by p, are open in addition to the radiative channel r. Widths for the decay of the compound state c will be denoted by a superscript c.

The probability of decay of c is the reciprocal of the lifetime τ^c.

Therefore the total width $\Gamma^c = \hbar/\tau^c$ is the decay probability expressed in energy units. According to probability theory the probability of one of a number of alternative events occurring is the sum of the separate probabilities. Therefore the total width is the sum of the partial widths for decay into each of the alternative channels.

$$\Gamma^c = \Gamma^c_r + \Sigma_p \Gamma^c_p. \tag{2.10}$$

If the total width Γ^c is less than the average spacing of levels D in the neighbourhood of the state c, discrete resonances will be observed in the scattering cross-section as a function of energy. Some resonances observed in a modern experiment with neutrons from a reactor are illustrated in Fig. 2.1. The extremely good energy resolution for such experiments is obtained by chopping the beam into short pulses by removing a neutron absorber for a short time. The time of flight of the neutrons is measured to determine their energy. The speed of a 10 eV neutron is of the order of 10^6 cm/sec and it is possible to resolve times of 10^{-9} sec with modern electronics.

As the energy increases, many more channels will open and the total widths of the resonances will increase until eventually they are larger than the spacings, which decrease with increasing energy. Channels at higher energies are those for the emission of different particles leaving the residual nuclei in various excited states, many of which have degeneracies in angular momentum projection. The cross-section at higher energies will exhibit fluctuations (which are not directly related to resonances) about some average value. Fluctuations in neutron cross-sections as a function of energy are shown in Fig. 2.2. Although there is no chance of predicting the resonance parameters it might be hoped that certain average values can be predicted. Most of the study of nuclear reactions has concentrated on this point. For the present, however, we will concentrate on the individual resonances.

The radial wave equation (1.10) for s-state neutrons in the exterior region is

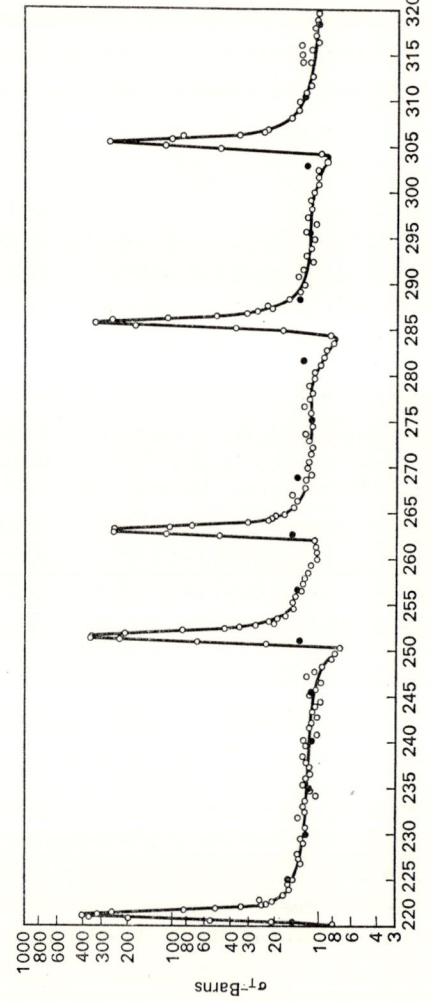

FIG. 2.1. The total cross-section for neutrons incident on Th^{232} as a function of incident energy between 220 eV and 320 eV (adapted from Stehn et al., 1965).

THE COMPOUND NUCLEUS

$$\frac{d^2 u_0}{dr^2} + k^2 u_0 = 0. \tag{2.11}$$

The scattering solution may be expressed in the form

$$u_0 = \tfrac{1}{2}(e^{-ikr} + \eta_0 \, e^{ikr}), \, r \geq R, \tag{2.12}$$

where
$$\eta_0 = e^{2i\delta_0}. \tag{2.13}$$

This form represents an ingoing spherical wave with unit amplitude and an outgoing spherical wave with amplitude η_0, called the reflection amplitude. If only the elastic channel is open $|\eta_0|^2 = 1$, and the only effect of the nucleus is a phase shift. If other channels are open some probability flux goes into these channels. Therefore less probability flux comes out in the entrance channel and $|\eta_0|^2 < 1$, so that δ_0 is a complex number. A complex phase shift is derived from a complex potential. At present we will not consider the mechanism for calculating η_0.

The behaviour of u_0 just outside the surface R will be described by the logarithmic derivative

$$L_0 \equiv R \left[\frac{du_0/dr}{u_0} \right]_R. \tag{2.14}$$

Using (2.12) we express η_0 in terms of L_0 as follows:

$$\eta_0 = \frac{L_0 + ikR}{L_0 - ikR} e^{-2ikR}. \tag{2.15}$$

We may obtain the s-state partial scattering amplitude from (1.20) using (2.13, 2.15):

$$f_0 = \frac{1}{2ik} \left[(e^{2ikR} - 1) - \frac{2ikR}{L_0 - ikR} \right]. \tag{2.16}$$

It may be seen from (2.12) that for hard sphere scattering, where u_0 vanishes at the channel radius R, the phase shift is kR. Therefore, by comparison with (1.20), the first term in the square bracket of (2.16) is called the hard sphere scattering amplitude or potential scattering amplitude. It varies slowly with energy. The second term in the square bracket is the resonance scattering amplitude.

28 NUCLEAR REACTIONS

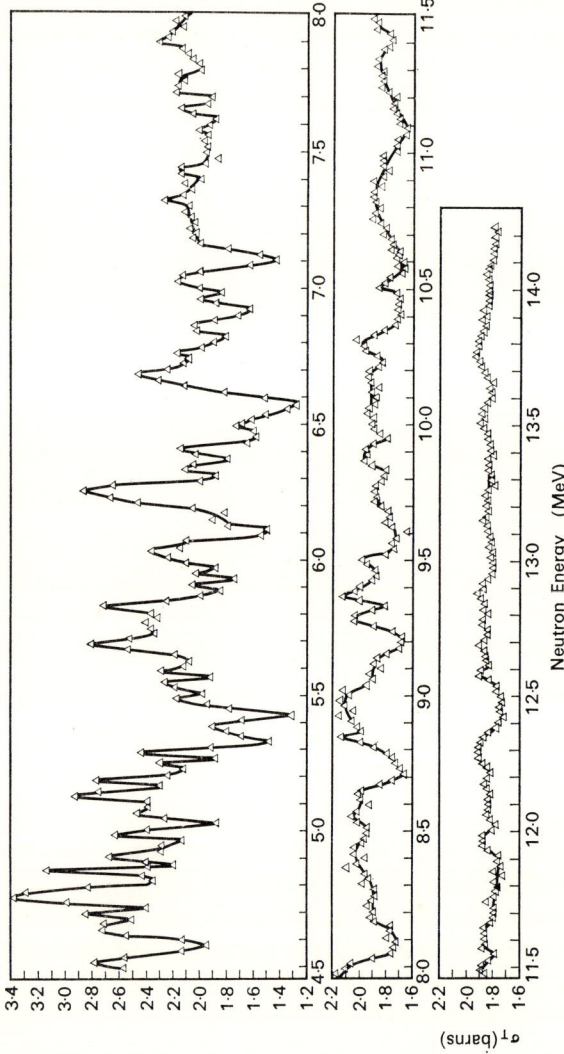

FIG. 2.2. The total cross-section for neutrons incident on Si^{28} in the energy range 4·5 MeV to 14·3 MeV (adapted from Carlson and Barschall, 1967).

The resonance scattering can be understood by considering the wave function $u_0(k, r)$ just inside the channel radius, where the wave number is K. It has the form of an ingoing spherical wave of unit amplitude and an outgoing wave with a phase change 2ζ determined by the many-body details of the interior interaction.

$$u_0(r) \simeq e^{-ikr} + e^{i(Kr+2\zeta)}$$
$$= 2e^{i\zeta} \cos(Kr + \zeta), r \gtrsim R. \qquad (2.17)$$

The internal conditions which determine ζ are not expected to be strongly dependent on the incident energy E. Certainly K is not. The logarithmic derivative may be obtained from (2.17).

$$L_0 \simeq -KR \tan(KR + \zeta) = -KR \tan z(E), \qquad (2.18)$$

where $z(E)$ is a smooth function of energy.

As the energy varies L_0 has zeros, for which the resonant scattering term in (2.16) is large. This provides us with the definition of the first of our resonance parameters, the resonance energy E^c of the compound state c. The definition is

$$L_0^c(E^c) \equiv 0. \qquad (2.19)$$

The resonance width Γ^c is defined by expanding $L_0(E)$ about its zero at E^c and neglecting all but the linear term in the power series.

$$L_0^c(E) \simeq [dL_0/dE]_{E^c}(E - E^c). \qquad (2.20)$$

We define Γ^c by

$$\Gamma^c \equiv -\frac{2kR}{[dL_0/dE]_{E^c}}. \qquad (2.21)$$

With these definitions the resonant scattering amplitude is

$$f_0^c = \frac{i\Gamma^c}{E - E^c + i\Gamma^c/2}. \qquad (2.22)$$

The square of this amplitude is proportional to the Breit–Wigner expression (2.9) for elastic scattering. The term involving the spherical scattered wave which was omitted by Breit and Wigner is the potential scattering term in (2.16). For elastic scattering the energy curve is like that in the 1931 calculation of Mott and like the experimental curves of Fig. 2.1. The resonant amplitude interferes destructively with the slowly varying potential scattering (which includes the Coulomb scattering in the case of charged incident particles) for $E < E^c$, and constructively for $E > E^c$. For energies far from E^c, L_0^c is of order KR and hence the resonant amplitude is much smaller than the potential scattering amplitude.

From (2.18, 2.19, 2.20, 2.21) we can obtain an estimate of Γ^c. At resonance the slowly varying function $z(E)$ is a multiple of π. We assume that $dz/dE \simeq \pi/D$, where D is the level spacing, so that

$$[dL_0/dE]_{E^c} = -KR[dz/dE]_{E^c} \simeq -KR\pi/D$$

and

$$\Gamma^c \simeq 2\frac{k}{K}\frac{D}{\pi}. \qquad (2.23)$$

Thus the level width is less than the spacing for $k \ll K$.

If channels p other than the entrance channel (for example the radiative channel r) are open, the phase ζ is complex with a positive imaginary part q. The linear term in the expansion of the logarithmic derivative is, by analogy with (2.20),

$$L_0^c(E) \simeq [\partial L_0/\partial E]_{E^c}(E - E^c) + [\partial L_0/\partial q]_{E^c}q$$
$$= [\partial L_0/\partial E]_{E^c}(E - E^c) - iqKR. \qquad (2.24)$$

The elastic scattering width is again given by (2.21). This time it is labelled by the entrance channel index s:

$$\Gamma_s^c \equiv -\frac{2kR}{[\partial L_0/\partial E]_{E^c}}. \qquad (2.25)$$

The reaction width Γ_R^c is defined by

$$\Gamma_R^c \equiv -\frac{2qKR}{[\partial L_0/\partial E]_{E^c}}. \qquad (2.26)$$

The total width is

$$\Gamma^c \equiv \Gamma^c_s + \Gamma^c_R. \qquad (2.27)$$

The resonant scattering amplitude is obtained by substituting (2.24, 2.25, 2.26, 2.27) in (2.16):

$$f^c_0 = \frac{i\Gamma^c_s}{E - E^c + i\Gamma^c/2}. \qquad (2.28)$$

We define the reaction cross-section by remembering that the reflection probability $|\eta_0|^2$ is now less than unity, the difference being the reaction probability.

The reaction cross-section is defined as

$$\sigma_R = \pi\lambdabar^2(1 - |\eta_0|^2). \qquad (2.29)$$

From (2.15, 2.24, 2.25, 2.26, 2.27) it is given by

$$\sigma_R = \pi\lambdabar^2 \frac{\Gamma^c_s \Gamma^c_R}{(E - E^c)^2 + (\Gamma^c/2)^2}. \qquad (2.30)$$

If only the channel r, for example, is open in addition to s, (2.30) corresponds to the Breit–Wigner formula (2.9). Note that Mott's formula (1.25) follows from the generalization of (2.29) to arbitrary l.

The reaction cross-section σ_R is proportional to the reaction width Γ^c_R, which, according to (2.10) is the sum of the partial widths for the excitation of the possible exit channels p. Therefore the cross-section for the excitation of the channel p from the compound state c is

$$\sigma_p = \pi\lambdabar^2 \frac{\Gamma^c_s \Gamma^c_p}{(E - E^c)^2 + (\Gamma^c/2)^2}. \qquad (2.31)$$

The physical interpretation of (2.30) is in terms of the fact that Γ^c_p represents the probability of the compound state c decaying into the channel p. The reciprocity theorem or time-reversal invariance is an essential property of nuclear systems. In this case it means that Γ^c_p is also the probability of formation of the compound state c if p is the entrance channel. Equation (2.31) states that the probability of a reaction in channel p resulting from the entrance channel s is given by the product of the probabilities for formation and decay of the compound nucleus. That is the formation and decay of the compound nucleus are statistically independent. This is an alternative statement of the compound nucleus model of Bohr.

The extension of (2.31) to higher values of l is simple. It is given by Blatt and Weisskopf. The extension to the case where more than one compound level c is important in the reaction is obtained by adding the reaction amplitudes for each value of c, provided the states c are discrete, that is provided the resonances do not overlap. The extension is the Breit–Wigner many-level formula. The reaction amplitude for exciting the channel p is given by analogy with (2.28). From (2.15, 2.29), the Breit–Wigner many level formula for s-state elastic scattering is

$$\eta_0 = \left[1 - i\Sigma_c \frac{\Gamma_s^c}{E - E^c + i\Gamma^c/2} \right] e^{2ikR}. \qquad (2.32)$$

The Breit–Wigner many level formula for a reaction other than elastic scattering involves the partial width amplitudes G_p^c whose absolute squares are the partial widths Γ_p^c. The resonance numerators are $G_p^c G_q^c$ for a (p, q) reaction.

The partial widths are quantities which depend on the external property k of the reaction. It is useful to separate out k by defining a new quantity γ_p^c called the reduced width, which depends only on the internal properties of the nucleus and the channel radius R. For s-state neutrons

$$\Gamma_p^c \equiv 2kR\,\gamma_p^c. \qquad (2.33)$$

The reduced width γ_p^c is obtained from the logarithmic derivative of the wave function for channel p, L_{0p}, by analogy with (2.21).

$$\gamma_p^c = -\,[dL_{0p}/dE]_{E^c}^{-1}. \qquad (2.34)$$

The reduced width amplitudes g_p^c, whose absolute squares are γ_p^c, are fundamental properties of the compound nucleus, since they are directly related to the logarithmic derivatives of the wave functions and are independent of the channel energies.

The significance of the reduced width amplitudes may be further understood by considering the definitions (2.7) of the partial widths according to Breit and Wigner, remembering that the widths defined there must be multiplied by $2\hbar$. Using the fact that the number of possible plane waves in the normalization volume V per unit frequency range is given by

$$1/\Delta v_s = 4\pi V/v\lambda^2, \qquad (2.35)$$

we have for channel p in the case where c is an s-state,

$$\begin{aligned}\Gamma_p^c &= 4\pi hV \mid A_p \mid^2/v\lambda^2 \\ &= 2k\mu V \mid A_p \mid^2. \end{aligned} \qquad (2.36)$$

The reduced width amplitude g_p^c is therefore proportional to the matrix element A_p. This depends on the overlap of the wave function of the compound state c with that of the channel p.

If the resonances in channel p overlap, the reaction amplitude at a particular energy near E^c is determined not only by one compound state c, but by others, c', as well. The amplitude corresponding to (2.32) then contains a double sum over c and c' including cross terms $G_p^c G_p^{c'}$ in the resonance numerators for elastic scattering in channel p, or $G_p^c G_q^{c'}$ for a (p,q) reaction. Since the scattering at a particular energy is not now determined by a simple product of partial widths, we cannot say that the formation and decay of the compound nucleus are independent.

It is just this assumption which is invalid at higher energies where resonances overlap. Thus Bohr's conjecture that the compound nucleus model is valid at all energies is expected to be incorrect.

A criterion for the validity of the independence assumption, which implies well-separated resonances, is the angular distribution for a reaction at a particular energy near E^c. The angular distribution is given by a particular partial wave whose l-value is equal to that of the resonant state c by angular momentum conservation. The cross-section is therefore proportional to $[P_l(\cos\theta)]^2$, which is symmetrical about $90°$. If resonances overlap, the cross-section is the square of a linear combination of different $P_l(\cos\theta)$, which is not in general symmetrical about 90°.

We now have an understanding of nuclear reactions in terms of resonance expansions of the reaction amplitude, for which s-wave elastic scattering of neutrons has served as an example. Of course for a higher energy reaction it is impossible to determine the resonance parameters and the expansion is useless for predicting results. It is necessary next to investigate relationships between resonance terms in the scattering amplitude with a view to deducing statistical properties of large groups of them.

2.3 Statistical Properties of Resonances

A very large number of compound states c and a much larger number of reduced width amplitudes are relevant to the reactions that can be induced by a particular probe and target. This is the statement in quantum mechanics of the complexity of a reaction first noted by Bohr. The complexity is analogous to that of a classical system to which statistical mechanics applies. The values of the many dynamical variables can in principle be determined, yet correct answers are obtained by assuming that the basic dynamical variables are random. The central limit theorem of statistics states that in the limit of an infinite sample independent variables are normally distributed with mean and variance determined by the nature of the variables. The normal distribution has a gaussian shape.

The basic variables of the theory of a nuclear reaction are the positions of the resonances (eigenvalues of the Hamiltonian of the compound system) and the reduced width amplitudes. The distributions of these quantities and the underlying statistical hypotheses necessary to determine them have been studied by Wigner, Thomas, Porter and others. Using low-energy neutrons from reactors it is possible to identify individual resonances and to obtain large enough samples to estimate the parameters of the distributions. Since the spin J and parity π of a state of the compound nucleus are good quantum numbers the Hamiltonian

THE COMPOUND NUCLEUS 35

matrix splits into disjoint submatrices for particular $J\pi$. We will discuss resonances with the same spin and parity.

The distribution of nearest-neighbour spacings between resonances is an example of a distribution that can be investigated experimentally. Wigner (1957) surmised that it is

$$P(x) = (\pi/2)x \exp(-\pi x^2/4), \qquad (2.37)$$

where x is the ratio of the spacing to the mean spacing. The correct distribution function was determined by Mehta (1960) with the assumption that the elements of the Hamiltonian matrix are normally distributed. It is almost the same as Wigner's formula. Examples of this distribution were produced by Porter and Rosenzweig (1960) who diagonalized large matrices with elements chosen according to the normal distribution hypothesis. The interesting thing about the spacing distribution is that levels tend to repel each other. The probability of small spacing is small. The distribution is confirmed by experiment.

The distribution of reduced width amplitudes was investigated by Porter and Thomas (1956). Their paper is reproduced in Part 2. The statistical assumption is that the basic quantities are normally distributed with zero mean. If the reduced width amplitudes are real then their absolute squares, that is the reduced widths which can be determined experimentally, are distributed as χ^2 with one degree of freedom. This distribution is

$$P_1(x) = x^{-1/2} \exp(-x/2), \qquad (2.38)$$

where x is the ratio of γ_p^c to the mean reduced width $\bar{\gamma}$, which is the variance of the reduced width amplitude distribution.

$$\bar{\gamma} = \overline{|g_p^c|^2}. \qquad (2.39)$$

The distribution of the sum of the squares of n independent, normally distributed quantities is χ^2 with n degrees of freedom. (A discussion of the χ^2 distribution is to be found for example in the book on the advanced theory of statistics by Kendall (1946). For our purposes it will suffice to state the results, when they are needed.)

If the reduced width amplitudes are essentially complex then $n = 2$ and the distribution is

$$P_2(x) = \exp(-x). \tag{2.40}$$

Porter and Thomas were able to show that the distribution for neutron scattering and radiation widths is $P_1(x)$, so that the reduced width amplitudes are essentially real and normally distributed about zero mean.

While these are the basic distributions, there are many more quantities whose distributions can be determined experimentally. An example is the fluctuation of the cross-section for a reaction as a function of energy in the energy region where the resonances overlap. The fact that large fluctuations are to be expected was noted by Porter and Thomas. We will consider elastic scattering as an example, but the same considerations apply exactly to other reactions.

The scattering amplitude may be written

$$f(E,\theta) = \frac{1}{2ik} \Sigma_l (2l+1) P_l(\cos\theta)$$
$$\left\{ C_l(E) + \Sigma_c \frac{R_l^c}{E - E^c + i\Gamma^c/2} \right\}, \tag{2.41}$$

where $C_l(E)$ is a slowly-varying potential scattering term and R_l^c is proportional to the partial width.

If the average total width Γ is many times greater than the average level spacing D, many resonances overlap and many terms contribute to the amplitude at a particular energy. We may write the differential cross-section as

$$d\sigma(\theta)/d\Omega = |\Sigma_l (\xi_l + i\eta_l) P_l(\cos\theta)|^2$$
$$= \xi(\theta)^2 + \eta(\theta)^2. \tag{2.42}$$

If a large number of resonances contribute to the sum, we make the assumption that ξ_l and η_l are normally distributed with the same variance. The distribution of the differential cross-section is χ^2 with two degrees of freedom, given by (2.40) with x equal to the ratio of the differential cross-section to the average differential cross-section.

The total cross-section is obtained by integrating (2.42) over θ. Because of the orthogonality of the Legendre polynomials the sum over l is now incoherent and we have the sum of the squares of $2l_0$ normally distributed quantities, where l_0 is the number of contributing partial waves. In general if there are N angular momentum substates due to the spins of the probe and the target the cross-section is distributed as χ^2 with $2N$ degrees of freedom.

$$P_{2N}(x) = \frac{N}{(N-1)!} (Nx)^{N-1} \exp(-Nx). \qquad (2.43)$$

These considerations have been discussed for example by Ericson (1963) and by Brink and Stephen (1963). They have been verified by many authors in cases where N is expected to be small.

An interesting application of the study of fluctuations is the determination of the average total width Γ of the resonances, which is related to the lifetime of the compound nucleus. Cross-sections for two different energies within the range $\pi\Gamma$ are correlated because the contributions come from the coherent addition of almost the same resonance contributions. However, at energies separated by much more than $\pi\Gamma$ the cross-sections are determined by entirely different resonance contributions. They are uncorrelated. A statistical analysis can be performed to determine an autocorrelation function which gives the correlation energy $\pi\Gamma$. Correlation energies for incident energies of several MeV are typically a few tens of kilovolts, corresponding to a compound nucleus lifetime of 10^{-19} sec.

2.4 The Statistical Model

So far the distributions we have discussed are those of quantities relative to their mean values over an energy range containing a large number of resonances. It is of great interest to know how the mean values themselves vary with energy for energy changes much larger than the averaging interval. They are not general properties of quantum systems, but are specific to particular compound nuclei. To determine them we need a mathematical model describing the mechanism of the reaction.

The statistical model or compound nucleus model is the one proposed by Bohr (1936) and Bohr and Kalckar (1937). Numerical discussions and examples are given for example by Bethe (1937). Its basic assumption is the independence of formation and decay of the compound nucleus. A compound nucleus, once formed, is considered as a complex system with heat energy distributed over the many degrees of freedom of the system. The process of emergence of a particle is analogous to that of evaporation. The energy distribution of evaporated neutrons is Maxwellian:

$$P(E) = E \exp(-E/\mathcal{T}), \tag{2.44}$$

where E is the energy of a particle of a particular type. The constant $\mathcal{T}$ is called the nuclear temperature. For incident energies above a few MeV, where there is a large amount of heat energy to be shared by the various degrees of freedom, the energy distribution of emitted particles in fact obeys the Maxwellian law for all but a small proportion of particles near the high energy end of the spectrum. Typical nuclear temperatures are of the order of 1 MeV. An example is shown in Fig. 2.3.

The thermodynamic analogy can be carried further to give a rough theory of the mean level spacing D whose reciprocal p is the mean level density. The probability of finding the compound nucleus in the energy range $(E, E + dE)$ is $\rho(E)\,dE$. The entropy is defined by

$$\mathcal{S} = \log \rho(E). \tag{2.45}$$

The average energy $\bar{E}$ of a system is a monotonically increasing function of temperature. Its derivative at $\mathcal{T} = 0$, the specific heat at $\mathcal{T} = 0$, vanishes according to the third law of thermodynamics. We assume that there is a power series expansion of $\bar{E}(\mathcal{T})$ in powers of $\mathcal{T}$, which must start with the quadratic term. Our rough theory assumes that

$$\bar{E} = a\mathcal{T}^2, \tag{2.46}$$

so that the entropy is given by

$$\mathscr{S} = \int dE/\mathscr{T}(E) = 2(a\bar{E})^{1/2} + \text{const.} \tag{2.47}$$

The level density is

$$\rho(\bar{E}) = C \exp[2(a\bar{E})^{1/2}]. \tag{2.48}$$

This dependence of ρ on $\bar{E}$ has been verified experimentally. For lighter nuclei a and C are of the order of 1 MeV^{-1}, while a increases to more than 10 MeV^{-1} and C decreases below 10^{-2} MeV^{-1}

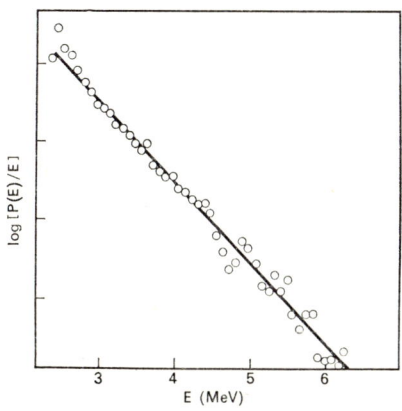

FIG. 2.3. The energy distribution of the final state for the inelastic scattering of 10 MeV protons on Zn^{64}. Log$[P(E)/E]$ is plotted against E on an arbitrary linear scale. The straight line corresponds to $\mathscr{T} = 0.933 \pm 0.018$ MeV. The parameter a of (2.48) has the value 5.61 ± 0.15 MeV^{-1} (adapted from Johnson and Hintz, 1967).

as the mass number increases. The increase of the level density with energy is one fact which must be explained by a model of nuclear structure. There are certain values of A, for example 208, near which level densities are anomalously small. This must also be explained by a model of nuclear structure.

A simple explanation, due originally to Bohr, of the increase of level density with energy is given by considering the nucleus as a collection of A harmonic oscillators with energy

$$\Sigma_{i=1}^{A} (n_i + 1/2)\hbar\omega_i.$$

For a low excitation energy only a few of the n_i are not equal to zero, so that the energy jumps are of the order of $\hbar\omega_i$. For high excitation many of the n_i are large so that there are many different sets of n_i which give an energy close to the same value. Each set of n_i represents an energy level.

A more detailed reaction model, which is based on the same statistical assumption of the independence of formation and decay of the compound nucleus, was proposed by Wolfenstein (1951) and by Hauser and Feshbach (1952). This model assumes that the total spin J and parity π of the system remain constant throughout the interaction and that the probability of a state of particular orbital angular momentum l and spin j occurring in any of the systems, probe, target, residual nucleus or emitted particle, is proportional to the number of magnetic substates which can be formed, except that it is weighted by a transmission coefficient $T_l(E)$ for each l and energy in either the initial or the final state. This means that if it is possible for a compound nucleus to form, its formation probability is the same for each possible quantum state of angular momentum, angular momentum projection, and parity. The reciprocity theorem makes exactly the same considerations hold for the decay of the compound system. The model is explained in the paper by Feshbach, Porter and Weisskopf in Part 2.

The reason why it is not certain that a compound nucleus can form is the possible reflection of the incident particle with a reflection amplitude η_l, defined by the analogue of (2.12) for general l. For large separation distance r,

$$u_l \sim (2i)^{-1} [\exp - i\,(kr - l\pi/2 - \alpha \log 2kr) \\ - \eta_l \exp i(kr - l\pi/2 - \alpha \log 2kr)], \quad (2.49)$$

where

$$\eta_l = e^{2i\,(\delta_l + \sigma_l)}. \quad (2.50)$$

The transmission coefficient is the probability of a particle not being reflected:

$$T_i(E) = 1 - |\eta_l|^2. \tag{2.51}$$

The Hauser–Feshbach model has become very useful in recent times when so much data on nuclear spectroscopy have been accumulated that it is possible to estimate the energy, spin and parity of every accessible state in some reactions (see for example Vogt, McPherson, Kuehner and Almquist, 1964). The transmission coefficients may be estimated with the aid of crude ideas about the potential inside a nucleus. However, once again, we will see that they may be explained by an accurate model for elastic scattering.

2.5 Mean Cross-sections: the Continuum Model

The most obvious mean quantity to be understood in the case of overlapping resonances is the mean cross-section. We will discuss it for the case of elastic scattering of s-wave neutrons. The energy region where resonances overlap is sometimes called the continuum, although we have seen that the cross-section exhibits fluctuations with correlation energies of tens of kilovolts at an incident energy of a few MeV.

The assumption of the independence of formation and decay of the compound nucleus will again be made in a modified form. As we have seen, the general argument for this assumption that the scattering amplitude depends on the product of partial widths is not valid for overlapping resonances. The assumption now becomes a model for the reaction rather than a generally valid statement. It is the principal assumption of the continuum model of Feshbach, Peaslee and Weisskopf (1947). Its justification is the fact that many compound states must be superposed to give the scattering amplitude at a particular energy. The contributions are considered to be random. We have seen that this assumption is well justified. In particular the signs and magnitudes of the reduced width amplitudes for resonances with particular quantum numbers are random. Since each compound state can give rise to many different decay modes this means that the decay modes also are random.

The reason that the resonances overlap is that many channels are open so that the total widths are the sums of many partial widths. This occurs at several MeV, where excited states of the nuclei involved in the reaction can take part. The assumption of the independence of formation and decay means that no emitted particles can contribute to an outgoing wave which is coherent with the incident wave. The entrance channel can be neglected since so many other channels contribute. Elastic scattering under this assumption is called compound elastic scattering.

The assumption is made in much of the literature that the potential and resonance scattering cross-sections are incoherent. It is sometimes justified by a time-dependent argument such as the following. (i) The incident beam may be considered as made up of wave packets whose width in time is the reciprocal of the energy definition of the beam. (ii) In order to discuss the mean cross-section we require poor energy resolution which corresponds to narrow wave packets in time. The scattered wave packet emerges long after the incident one has struck the nucleus, since the compound nucleus life-time is long. If the wave packets are sufficiently narrow they will not overlap, hence the scattered beam will be incoherent with the incident beam.

The first of these arguments is quite incorrect. Consider scattering from a plane wave state $\exp i\mathbf{k}_i \cdot \mathbf{r}$ to another $\exp i\mathbf{k}_f \cdot \mathbf{r}$. The scattering amplitude may be considered simply as the matrix element of a scattering operator T for incident and outgoing wave numbers k_i and k_f respectively in directions $\hat{\mathbf{k}}_i$ and $\hat{\mathbf{k}}_f$.

$$f(k_f, k_i) = \langle e^{i\mathbf{k}_f \cdot \mathbf{r}} \mid T(r, r') \mid e^{i\mathbf{k}_i \cdot \mathbf{r}} \rangle. \tag{2.52}$$

Each wave packet is a superposition of plane waves with a form factor $A(k, \delta)$, whose width in k space is δ. Since we are making an energy-time argument it will be sufficient to consider one dimension. The outgoing wave packet for example is

$$\psi_f(k_f, \delta_f, r_f, r) = \int dk\, A_f(k_f, \delta_f)\, e^{ik(r-r_f)}. \tag{2.53}$$

This wave packet is centred on a wave number k_f with width δ_f. It starts at a position r_f at $t = 0$.

THE COMPOUND NUCLEUS   43

The scattering amplitude for wave packets is

$$f(k_f, k_i, r_f, r_i) = \int dk \int dk' \, A_f^*(k_f, \delta_f) \, A_i(k_i, \delta_i) \, e^{i(kr_f - k'r_i)}$$
$$\times \int dr \int dr' \, e^{-ikr} T(r, r') \, e^{ik'r'}. \qquad (2.54)$$

In a scattering experiment time is not resolved. Even with the best electronics time resolution of only 10^{-10} sec is possible. This corresponds to an energy resolution of about 10^{-5} eV which is essentially zero for nuclear physics. Therefore we have no knowledge of r_f and r_i and must integrate over them giving equal weight to all values.

The cross-section may be written

$$\int dr_f \int dr_i \, |f(k_f, k_i, r_f, r_i)|^2$$
$$= \int dr_f \int dr_i \int dk \int dk' \int dr \int dr' \, A_f^* A_i e^{i(kr_f - k'r_i)} \, e^{-ikr} T(r, r') \, e^{ik'r'}$$
$$\times \int dk'' \int dk''' \int dr'' \int dr''' \, A_f A_i^* \, e^{-i(k''r_f - k'''r_i)} \, e^{ik''r''} T(r'', r''') \, e^{-ik'''r}$$
$$= \int dk \int dk' \, |A_f|^2 \, |A_i|^2 \, |\langle e^{ikr} | T(r, r') \, e^{ik'r'} \rangle|^2. \qquad (2.55)$$

Thus the scattering of wave packets in the absence of time resolution is equivalent to a series of incoherent experiments for scattering of plane waves with energy distributions given by the energy spreads of the wave packets. In other words particles of different energies in a beam may be separated out (for example by magnetic analysis), separate scattering experiments performed with each energy component, and the beams re-aligned, without affecting the observed cross-section.

If argument (ii) is made with the reservation that it is instructive to consider what would happen in principle if sufficient time resolution could be obtained, then the theory of scattering of wave packets must be worked out in detail. This was done for elastic scattering by Dodd and McCarthy (1964), who showed that the interference term between the potential and resonance scattering for isolated resonances does in fact decrease as the wave packet narrows in time, but that the resonance cross-section decreases at the same rate. For overlapping resonances, however, the scattered beam comes out on the average at the same time as the potential scattering.

The assumptions of the continuum theory mean that there is a purely ingoing wave at the channel surface. In terms of the crude model for s-wave neutrons used at the beginning of section 2.2, the wave function just inside the channel radius R is

$$u_0(r) = e^{-iKr}, r \lesssim R. \tag{2.56}$$

Hence $\eta_0 = 0$ and (2.14) gives for the logarithmic derivative

$$L_0 = -iKR. \tag{2.57}$$

The absorption cross-section is

$$\sigma_R = \pi \lambdabar^2 \frac{4kK}{(k+K)^2}, \tag{2.58}$$

and the scattering amplitude is given by (2.16, 2.57):

$$f_0 = \frac{1}{2ik}\left[(e^{2ikR} - 1) + \frac{2ik}{k+K}\right]. \tag{2.59}$$

The scattering cross-section is a smoothly varying function of energy, approaching πR^2 as the energy increases.

For higher energy neutrons where $kR \gg 1$ we may consider the nucleus as a circular hole in the wave front, since particles which come within the radius R are absorbed. We neglect the fact that there is an uncertainty of order $\lambdabar$ in determining whether a particle with angular momentum $l = kR$ is inside or outside R. The angular distribution may be computed in the Fraunhofer approximation in which a space-dependent phase factor is integrated over the projection of the nuclear volume onto the plane perpendicular to the incident beam. The scattering angle is assumed to be small.

$$f(\theta) \cong \frac{ik}{2\pi} \int\int dA\, e^{-ik_f \cdot r}$$

$$\cong \frac{ik}{2\pi} \int_0^{2\pi} d\phi \left[\int_0^R r\, dr\, e^{-kr\theta \cos\phi} \right]. \quad (2.60)$$

The cross-section is obtained in terms of cylindrical Bessel functions $J_m(z)$ defined by

$$2\pi i^m J_m(z) = \int_0^{2\pi} d\phi\, e^{iz\cos\phi} \cos(m\phi). \quad (2.61)$$

We have for the differential cross-section the black disc formula

$$d\sigma/d\Omega = (kR^2)^2\, [J_1(kR\theta)/kR\theta]^2. \quad (2.62)$$

This formula was first derived for neutron scattering by Bethe and Placzek (1937). It is a diffraction pattern with a large peak at small angles and subsequent peaks decreasing rapidly in intensity.

The angular distribution is certainly not symmetrical about 90° so it is inconsistent with the Bohr assumption, as expected for higher energies. As was discussed at the beginning of this section, the key assumption $\eta_l = 0$ is not derived from the Bohr assumption, but from a random channel argument. However, although we assume that no particles from the compound nucleus return to the entrance channel, this does not mean that there is no elastic scattering. The removal of particles from the beam causes diffraction scattering. As the energy increases the total cross-section for uncharged particles approaches $2\pi R^2$. This includes πR^2 for diffraction scattering and πR^2 for the reaction cross-section. The latter is obtained as a simple generalization of (2.29) for energies so high that many partial waves contribute:

$$\sigma_R = \pi \lambdabar^2 \Sigma_l (2l + 1)\, (1 - |\eta_l|)^2, \quad (2.63)$$

where
$$\eta_l = 0,\, l \leqslant kR,$$
$$\eta_l = 1,\, l > kR. \quad (2.64)$$

Putting $l = kr$ and approximating the sum by an integral we have

$$\sigma_R = (\pi/k^2)\int_0^{kR} d(kr)\,(2kr+1)$$

$$\cong \pi R^2. \tag{2.65}$$

Although the Bohr assumption does not hold for elastic scattering in the continuum model because of diffraction scattering, there is nothing in the model to say that it does not hold for the non-elastic channels through which the compound nucleus may decay. For some time many physicists expected non-elastic reactions to have angular distributions which are symmetrical about 90°, because of the random signs of the reduced width amplitudes, even though the resonances overlap.

2.6 Experiments and Their Analysis up to 1952

Angular and energy distributions for isolated resonances are completely understood in terms of the resonance parameters. For the region of overlapping resonances the continuum model could be tested by such experiments. Until the early 1950's very few experiments on angular or energy distributions for particular reactions (that is ones in which a single exit channel is experimentally isolated) were performed owing to experimental difficulties. In the earliest experiments events were observed by counting flashes of light caused by charged particles when they hit a scintillating screen. The work was done by eye in a darkened room and required considerable personal skill and practice. Charged particles do not travel far in air or other matter. In fact their ranges can be observed and calibrated so that their energies can be determined if the nature of the particle is known. Early visual techniques also included stereo photographs of cloud chamber tracks and tracks in photographic emulsions. Such techniques are still used.

The first device that lent itself to automatic data recording was the Geiger–Müller counter. We will give a general description of automatic data recording so that some of the difficulties and advances may be understood. The history of nuclear physics is

THE COMPOUND NUCLEUS 47

bound up very closely with the history of technological advances. Energies of charged particles are determined from the intensity of the radiation which they produce when stopped by matter. For neutrons detectors use the radiation from the interaction of secondary knocked-on particles. The knock-on cross-section is small, so these detectors are inefficient.

The radiation is turned into an electric pulse by means of the photo-electric effect. The pulses are amplified and the pulse heights are sorted and analysed automatically to give a digital representation of the number of pulses in a certain energy interval. Fast digital equipment was not developed until the late 1950's so that it was a long and difficult job to identify particles and measure their energies at a set of different scattering angles.

The speed of an experiment is limited by several factors. The resolving time of the counting and recording equipment is finite. This means that the maximum useful intensity of the beam is limited by the restriction that only one particle must be counted at a time. The number of data recording channels is limited by the capacity of the computing equipment available. The intensity of the beam is limited by the characteristics of the accelerator.

Early angular distribution experiments used beams of natural α-particles which were available with high enough energy for significant effects. Secondary beams, for example neutron beams, could be produced at higher energies, but they were not monoenergetic, so that nuclear states could not be identified. Their intensity was prohibitively low. We have already mentioned the use of early α-particle angular distributions in determining nuclear radii. As early as 1921, Chadwick and Bieler (1921) showed that the cross-section at backward angles for α, p scattering was many times the Rutherford value. Riezler (1932) performed systematic α-particle scattering experiments on light nuclei at different energies and angles, finding for example a resonance in α, C^{12} scattering of width 1 MeV centred at 5 MeV, which can be explained by a one-body theory like that of Mott (1931).

These experiments were analysed by means of a phase shift analysis, where values of the lower nuclear phase shifts δ_l are adjusted to fit the data. One of the first such analyses was that of Taylor (1932), who found that α, p angular distributions at very low energy could be fitted by postulating a value for δ_0 only. Wheeler (1937) used phase shifts up to δ_4 to describe α,α scattering.

Even up to the early 1950's sufficiently detailed energy and angular distributions of neutrons had not been obtained to provide a thorough test of the continuum model. The total cross-section had been measured at isolated energies and used to determine R from the approximate formula $2\pi R^2$. Values of $r_0 = R/A^{1/3}$ were approximately consistent with ideas developed from α-particle scattering, but variations of 20 per cent were not uncommon.

An important technological development was that of the synchrocyclotron in 1947 by the group at the Lawrence Radiation Laboratory, Berkeley. This enabled the relativistic barrier on the energy of accelerated particles to be broken. In a non-relativistic cyclotron the circulation frequency of the particles is constant. The radius of the orbit increases with the energy increase at each acceleration in exactly the right manner for the accelerating radio-frequency field to be kept at fixed frequency. The particle motion is isochronous. The barrier is due to the fact that at relativistic energies the kinetic energy no longer increases as $\frac{1}{2}mv^2$, so that the circulation frequency of particles is no longer the same as for non-relativistic energies and the particles become out of phase with the accelerating potential. The energy limit for protons is about 20 MeV. As early as 1937 it was shown by Thomas (1937) that an azimuthally-varying magnetic field would correct this difficulty. However, the computational problem involved in solving the orbit equations in a complicated field was too great for such a machine to be designed.

It was discovered independently by Veksler (1945) and McMillan (1945) that a short burst of particles would remain stably bunched throughout their passage through the cyclotron. Therefore the accelerating frequency could be varied with the radial position of the bunch in such a way as to maintain an accelerating potential at high energies.

Total neutron cross-sections at 90 MeV incident energy were

measured by Cook, McMillan, Peterson and Sewell (1949). They were analysed by Fernbach, Serber and Taylor (1949) by means of a variant of the black nucleus model, which was due originally to Bethe (1940). It was assumed that the de Broglie wavelength $\lambdabar = 0.5$ fm was small enough for semi-classical approximations to be made so that the path of a neutron in nuclear matter had some meaning. Instead of being certainly absorbed, it was assumed that the neutron had a finite mean free path which could be calculated from the nucleon–nucleon cross-section. The value obtained was $\lambdabar = 4$ fm. At the same time it was assumed that there was a real potential which produced an index of refraction in nuclear matter. This paper is reproduced in Part 2. A consistent value of r_0 for different targets was found, $r_0 = 1.37$ fm.

The significance of a mean free path λ may be seen by its effect on the wave number of a plane wave travelling through a medium with wave number k.

$$\psi(r) = e^{ikr} e^{-r/2\lambda} = e^{i(k+i/2\lambda)r}. \tag{2.66}$$

A particle travelling in an absorbing medium has a complex wave number and a complex index of refraction, which result from a complex potential. In the language of scattering theory (2.49, 2.50) the phase shift δ_l is complex and the magnitude of the reflection amplitude η_l is less than unity so that the transmission coefficient $T_l(E)$ (2.51) is somewhere between 0 and 1 for each partial wave that contributes to the reaction.

It is interesting to consider the solution (2.66) of the Schrödinger equation for a particle in a uniform medium where the potential is $V + iW$. We will assume that $W/(E-V)$ is small. The Schrödinger equation is

$$(\nabla^2 + k^2)\psi = 0, \tag{2.67}$$

where $$k^2 = \frac{2\mu}{\hbar^2}(E - V - iW),$$

$$k \simeq \left[\frac{2\mu}{\hbar^2}(E-V)\right]^{1/2}\left(1 - \frac{i}{2}\frac{W}{E-V}\right). \tag{2.68}$$

The imaginary part of the wave number (2.68) may be compared with that of (2.66) to obtain

$$W = - \left[\frac{\hbar^2}{2\mu}(E - V)\right]^{1/2} \bigg/ \lambda. \qquad (2.69)$$

The imaginary part of the potential W is negative for absorption and inversely proportional to the mean free path. For fixed W the mean free path increases with increasing energy.

In this chapter we have seen that the second twenty years of nuclear reaction studies was marked by an understanding of isolated-resonance phenomena and a series of conjectures about reactions in the region of overlapping resonances which could not be resolved because of experimental difficulties. The conjecture whose apparent validity was supported by nearly all experiments with neutrons was the strong coupling compound nucleus model of Bohr, which led to the continuum model. We have seen that the resonance expansion of a reaction amplitude confirms Bohr's ideas for isolated resonances and that it leads because of random reduced width amplitudes to the continuum model. The essential point of the compound nucleus philosophy is that reactions are so complicated that it is impossible to understand them mechanistically in terms of the motion of particles in potentials. The essential complication leads to statistical models which work very well in their range of validity.

However, throughout the second twenty years there were experiments, unsupported by calculations as detailed as the compound nucleus calculations, that could be explained on the basis of particle motion in a potential. We have noted some of the resonances of approximately 1 MeV width observed in early α-particle elastic scattering and the success of Fernbach, Serber and Taylor twenty years later in using what was essentially a complex potential model to describe reactions with high-energy neutrons.

Even the continuum model predicts certain features of potential scattering for elastic scattering experiments, but for non-elastic reactions this does not apply. Since 1935 one non-elastic reaction had been known whose total cross-section could be explained on the basis of particle motion. The (d,p) reaction was explained by Oppenheimer and Phillips (1935) as stripping. This means that the proton of the deuteron (which is separated from the neutron on the average by about 4 fm) is repelled by the Coulomb force while the neutron is not. The nuclear forces are enough to break up the deuteron, leaving the proton in a scattering state while the neutron is captured. At energies where the proton can penetrate the Coulomb barrier a (d,n) reaction is also possible. Such a reaction occurs in the time (10^{-22} sec) for a particle to cross the nucleus. No long-lasting compound state is formed. It was argued by Serber (1947) that high-energy reaction products from high-energy collisions would be largely described by the immediate emergence of a particle after the first elementary collision, once again in a short time, so that the Bohr collision mechanism would be invalid. His arguments were based on a survey of (d,n) cross-section data from experiments with the Berkeley synchrocyclotron (Serber, 1947a).

Attempts to understand the structure of bound nuclei were divided in the same way as those for reactions. The model of strong coupling and complicated motion was the liquid drop model of Bohr and Kalckar (1937). This model was supported by information, obtained by studying fission, about the stability of a nucleus against deformation. It did not, however, predict the correct energies of excited states of nuclei, which it should explain as collective surface vibrations. The single-particle model leads to periodic properties of nuclei as the mass number A is varied, just as it does for atoms. Properties such as the exceptional stability of He^4, O^{16} and certain heavier nuclei were found, but again could not be predicted in detail. Nuclear structure models will be discussed in Chapter IV.

Theoretical ideas current during the second twenty years are summarized and developed in three remarkable review articles by Bethe and Bacher (1936), Bethe (1937) and Bethe and Placzek (1937).

In spite of the strong suggestions of the validity in certain circumstances of single-particle models for both reactions and bound nuclei, the compound nucleus model was the only one with a sound theoretical basis, namely the resonance description originated by Breit and Wigner.

III

The Optical Model

SINCE the resonance expansion of the amplitude for a reaction follows perfectly generally from quantum mechanics, any explanation of phenomena averaged over resonances must be derivable from it. The first such attempt was the continuum model which predicted neutron total cross-sections of the order of $2\pi R^2$ which decreased monotonically to this value with increasing energy.

In 1952 neutron scattering experiments at closely-spaced intervals between $0·05$ MeV and $3·2$ MeV were performed by the Wisconsin group directed by Barschall. The preliminary report by Barschall (1952) is included in Part 2. The energy distributions are reviewed in detail by Miller, Adair, Bockelman and Darden (1952) and later angular distributions by Walt and Barschall (1954). Instead of smoothly decreasing total cross-sections, broad maxima in energy were observed which moved towards higher energy with increasing A. Their widths were of the order of 1 MeV, suggesting that they could be described by a one-body potential scattering model. This suggestion was strengthened by the fact that features of the energy distributions changed gradually with A, indicating that they depended on gross nuclear properties such as the radius rather than details of the structure of particular nuclei. In fact the experimental energy resolution was not sufficient to observe the basic fluctuations of the cross-sections. What was being measured was exactly the mean cross-section discussed in section 2.5.

3.1 The Explanation of Feshbach, Porter and Weisskopf

The explanation of potential scattering phenomena in terms of the resonance theory of reactions was given in a classic paper by Feshbach, Porter and Weisskopf which is reproduced in Part 2. The reader is expected to be able to follow it with the background provided in Chapter II. The effects of this work are far-reaching. It shows that it makes sense in terms of basic quantum mechanics to consider a particle as being capable of moving in a nucleus without necessarily changing the quantum state of the nucleus. Therefore, the wave function of such a particle may form the basis of a mechanistic theory of reactions.

It was shown by Feshbach, Porter and Weisskopf that the total cross-section σ may be divided into the shape elastic (or potential scattering) cross-section σ_{se} and the cross-section for formation of a compound nucleus σ_c. The latter cross-section is made up of the compound elastic cross-section σ_{ce}, due to the absorption of particles from the entrance channel and their re-emission in the entrance channel, and the reaction cross-section σ_r. It is generally assumed that σ_{ce} and σ_{se} are incoherent because of the random signs of the resonance contributions to the average scattering amplitude which make the cross terms vanish on squaring the amplitude. A physical way of understanding the difference between σ and σ_{se} is that absorption removes the particle from coherent interaction with the elastic particles, no matter what happens to it after absorption. However, we must be cautious about too-literal interpretation of time-dependent arguments in a time-independent experiment. The argument leading to equation (2.55) shows that all elastic scattering is basically coherent.

The average partial cross-sections $\sigma_{se}^{(l)}$ and $\sigma_c^{(l)}$ are given by

$$\sigma_{se}^{(l)} = \pi\lambda^2(2l+1) \mid 1 - \bar{\eta}_l \mid^2,$$
$$\sigma_c^{(l)} = \pi\lambda^2(2l+1) [1 - \mid \bar{\eta}_l \mid^2]. \quad (3.1)$$

These are the cross-sections derived from a reflection amplitude $\bar{\eta}_l$ for an average problem, where $\bar{\eta}_l$ varies slowly with energy (see equations (1.22), (2.29)). The average problem or gross structure problem is taken to be the scattering of a neutron by a complex potential whose real part reflects the size of the nucleus and the collective effect of all the nucleon–nucleon forces, and whose imaginary part is proportional to the probability of a particle being absorbed (see equation (2.69)). It is called the optical model.

The considerations leading to (3.1) depend on the statistical properties of the parameters of the resonance terms in the average *amplitude* and are perfectly general. In establishing quantitative connections between the average partial width Γ_0 for the entrance channel, the average spacing D and the reaction cross-section for the gross structure problem, the approximation is made that $\Gamma_0 \ll D$. In the total cross-section the contributions from different values of l add incoherently. Therefore the restriction $\Gamma_0 \ll D$ applies only to resonances with the same spin and parity.

We will establish the connection between the average resonance parameters and the gross structure problem in the case of non-overlapping resonances for s-wave neutrons where η_0 is given by (2.32). The optical model potential is *defined* to be that potential which reproduces the average reflection amplitude $\bar{\eta}_0$. The average partial cross-section is $\overline{\sigma^{(0)}}$, where the bar denotes an average over an energy interval I which includes many resonances, but which is small compared with the width of the gross structure resonances.

$$\overline{\sigma^{(0)}} = \pi \lambda^2 \ \overline{\mid 1 - \eta_0 - \bar{\eta}_0 + \bar{\eta}_0 \mid^2}$$
$$= \pi \lambda^2 \ \{\overline{\mid \bar{\eta}_0 - \eta_0 \mid^2} + \mid 1 - \bar{\eta}_0 \mid^2 + 2 \operatorname{Re} \overline{(\bar{\eta}_0 - \eta_0)} (1 - \bar{\eta}_0)^*\} \tag{3.2}$$

This clearly divides into $\overline{\sigma_{se}^{(0)}}$ and $\sigma_c^{(0)}$ just as in equations (3.1) if

$$\overline{\eta_0 - \bar{\eta}_0} = 0. \tag{3.3}$$

Equation (3.3) is the definition of the optical model amplitude η_0. Using the form (2.32) for η_0 and $\bar{\eta}_0$ we have

$$\eta_0 - \bar{\eta}_0 = i\, e^{2ikR} \left[\Sigma_m \frac{\Gamma^m}{E - E^m + i\Gamma^m/2} - \Sigma_c \frac{\Gamma_0^c}{E - E^c + i\Gamma^c/2} \right], \tag{3.4}$$

where the first term in the bracket refers to gross structure resonances with parameters E^m, Γ^m. The averaging operation leaves this

56 NUCLEAR REACTIONS

term constant. Note that the eigenvalues E^m of the complex potential problem are complex. The widths of the gross structure resonances therefore are comparable with the imaginary potential.

The average of (3.4) is zero according to (3.3). We have

$$\left\langle \Sigma_c \frac{\Gamma_0^c}{E - E^c + i\,\Gamma^c/2} \right\rangle = \Sigma_m \frac{\Gamma^m}{E - E^m + i\,\Gamma^m/2}. \quad (3.5)$$

(Here the average is denoted by brackets for convenience of notation.) The real and imaginary parts of the right-hand side of (3.5) are $R(E)$ and $Q(E)$ respectively where

$$Q(E) = \operatorname{Im} \left\langle \Sigma_c \frac{\Gamma_0^c}{E - E^c + i\Gamma^c/2} \right\rangle$$

$$\equiv \pi \Gamma_0 / D. \quad (3.6)$$

Equation (3.6) defines the average partial width and separation Γ_0 and D. It is equivalent to equation (2.15) of Feshbach, Porter and Weisskopf, where it is explained. We now have for $\bar{\eta}_0$

$$\bar{\eta}_0 = e^{2ikR} [1 - i(R + iQ)], \quad (3.7)$$

and for $|\bar{\eta}_0|^2$, using (3.6) for Q,

$$|\bar{\eta}_0|^2 = 1 - 2\pi\Gamma_0/D + [R^2 + \pi^2\Gamma_0^2/D^2]. \quad (3.8)$$

For small values of Γ_0/D and R we have the following relation between the s-wave partial absorption cross-section in the optical model and the s-wave ratio of mean level width to mean spacing.

$$\sigma_c^{(0)} = \pi\lambda^2(1 - |\bar{\eta}_0|^2)$$

$$= 2\pi^2\lambda^2\, \Gamma_0/D. \quad (3.9)$$

Note that $\Gamma_0 \leqslant D/2\pi$, since $\sigma_c^{(0)} \leqslant \pi\lambda^2$.

The relation (3.9) provides a test of the optical model. Since the quantity Γ_0/D is proportional to k it may be determined at energies where individual resonances can be observed and averaged, and easily transformed to different energies. It is sometimes called the strength function. The optical model potential which is chosen to fit averaged energy and angular distributions must also fit the resonance properties. As a function of A, $\sigma_c^{(0)}$ rises and falls with the gross structure resonances. Fits to this structure with different optical models are shown in Fig. 3.1.

The variation of Γ_0/D as a function of A at very low energies actually provides a better illustration of the gross structure resonances than the energy variation of the total cross-section. At energies of a few MeV, higher partial waves contribute to the cross-section and complicate the effect.

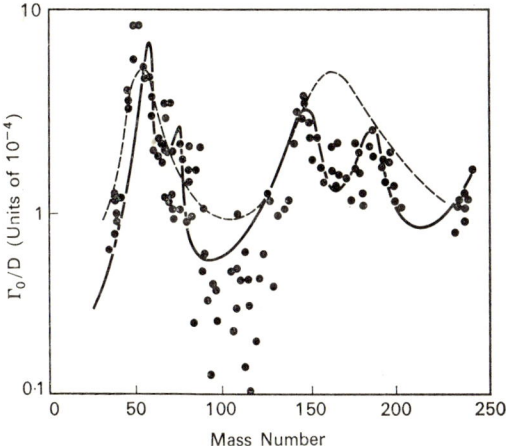

FIG. 3.1. The quantity Γ_0/D, normalized to an incident energy of 1 eV, plotted against mass number. The dotted curve is computed from the non-local optical model (Perey and Buck, 1962). The solid curve was computed by Chase, Wilets and Edmonds (1958) from a generalized optical model including low-lying collective states (see Chapter V).

3.2 Elastic Scattering Experiments and Optical Model Fits

The development of the optical model coincided with the development of experimental techniques to the point where angular distributions for the elastic scattering of protons, neutrons, deuterons, α-particles and heavier ions could be measured at different energies. Soon afterwards electronic computers were developed to the stage where optical model scattering amplitudes at energies involving up to 20 or more partial waves could be calculated at sufficient speed to allow fitting of angular distributions by adjusting parameters. Parameters were first adjusted by trial and error, but fitting codes were soon developed which chose parameters so as to minimize $\chi^2 = \Sigma_i(T_i-E_i)^2/E_i^2$, where T_i is the theoretically predicted number for the ith piece of data, E_i.

Accelerators in use at this time were non-relativistic cyclotrons, which could produce 20 MeV protons or 40 MeV α-particles, synchrocyclotrons for proton energies of 100 MeV or over, and Van de Graaff machines for low energies. An important new machine was the proton linear accelerator, the first of which was developed at Berkeley for accelerating protons to 31 MeV. Such machines use a radio-frequency potential to accelerate particles in small steps in a straight line, so that the phase difficulty encountered with cyclotrons is not present. A linear accelerator which accelerated protons to 10 MeV, 40 MeV or 68 MeV was built at Minnesota. These machines were invaluable for filling in the range above 20 MeV, and experiments performed with them confirmed the value of the optical model in different energy ranges and throughout the periodic table.

From an experimental point of view linear accelerators and synchrocyclotron have important disadvantages not present in Van de Graaff machines. It is difficult or impossible to vary the beam energy in small steps. Also they have a very small duty cycle (less than about 10 per cent). This means that even if the maximum beam intensity capable of being used by the counting equipment could be produced in short bursts, the beam would actually be hitting the target less than 10 per cent of the time so that experiments would be very slow. The small duty cycle of the synchrocyclotron is inherent in the idea of the machine, although it can be corrected to some extent after the acceleration process in modern machines. For a linear accelerator it is necessary in order to allow dissipation of the large quantities of heat that are generated by eddy currents in the tank that surrounds the accelerating tube and acts as a wave guide for the radio-frequency power.

THE OPTICAL MODEL   59

The tandem Van de Graaff was developed in the 1950's. This machine accelerates negative ions to a high-voltage terminal where the electrons are stripped off so that positive ions can be accelerated back to ground potential in the last half of the machine. It doubles the energy available from an electrostatic generator. Tandem Van de Graaffs of various sizes have now been developed which can accelerate protons to about 12 MeV, 15 MeV or 20 MeV. Such machines have very fine energy adjustment with resolution of 10 keV or better, which is better than that of the counting equipment. They have continuous beams and just enough intensity for ordinary scattering experiments. Their limitation is the potential which can be supported on the high-voltage terminal. So far this is not much above 10 MV. In a few machines a third stage of acceleration is used. A third stage with the largest machines would make 30 MeV possible.

The development of fast digital computers in the middle 1950's produced another major advance in accelerator design. It became possible to design an azimuthally-varying-field isochronous cyclotron. Such machines have no theoretical energy limit. Beam currents are so large that good energy resolution, if needed, can be obtained by magnetic analysis. Their duty cycle is that of the radio-frequency accelerating power supply which is roughly at the limit of the time resolution of counting equipment so that the effective duty cycle is 100 per cent. It is difficult and expensive to extract positive ions over large energy ranges and to alter the energy of the machine, although both these things are done with certain machines. These difficulties disappear when negative ions are accelerated, since the ions need merely to be stripped and are then extracted automatically by the cyclotron magnetic field. The radial position of the stripper determines the beam energy, which is therefore easily adjustable. The theoretically limiting factor on a negative ion machine is $v \times B$ where v is the particle velocity and B is the cyclotron magnetic field. This electric field dissociates the negative ions by the Stark effect. High energy machines have small magnetic fields and very large radii.

With the machines now at our disposal it is possible to obtain very accurate data and determine up to ten parameters describing optical model potentials.

The potential of Feshbach, Porter and Weisskopf had three parameters. The potential was assumed to be constant at a value $V + iW$ inside the square well radius $R = r_0 A^{1/3}$ and zero outside. Woods and Saxon (1954) found that very satisfactory fits to proton angular distributions at 20 MeV could be obtained with a potential that had a diffuse surface. The four-parameter Woods–Saxon potential $U(r)$ is

$$U(r) = -(V + iW) f(r, r_0, a)$$
$$= -(V + iW) \{1 + \exp[(r - r_0 A^{1/3})/a]\}^{-1}. \quad (3.10)$$

The factor $f(r, r_0, a)$ is the Woods–Saxon form factor which is nearly 1 inside the nucleus, is $0 \cdot 5$ at $r_0 A^{1/3}$, and falls exponentially to zero with large r. The diffuseness parameter is a. The distance between the points where $f = 0 \cdot 1$ and $f = 0 \cdot 9$ is $4 \cdot 4a$.

The four-parameter model was shown by several authors to provide good fits to differential cross-sections for momentum transfers up to about 1 fm^{-1}, but it could not be made to fit backward scattering. Typical fits are shown in Fig. 3.2.

Both proton and neutron beams were discovered by means of double scattering experiments to suffer large polarizations at certain angles. This is due to a spin–orbit coupling term in the potential which is often represented by

$$U_S(r) = \boldsymbol{\sigma} \cdot \boldsymbol{l} \left(\frac{\hbar}{M_\pi c^2}\right)^2 V_S \frac{1}{r} \frac{d}{dr} f(r, r_{0S}, a_S). \quad (3.11)$$

Here $\boldsymbol{\sigma}$ is the projectile spin operator and $\boldsymbol{l}$ its orbital angular momentum operator. The potential is normalized by the factor containing the pion mass M_π. The factor is peaked at the radius $r_{0S} A^{1/3}$. The radial wave equation corresponding to (1.10) is

$$\frac{d^2 u_\mp}{dr^2} + \left\{ k^2 - \frac{2\mu}{\hbar^2} \left[U_c(r) + \frac{\hbar^2}{2} \binom{l}{-l-1} U_S(r) \right] \right.$$
$$\left. - \frac{l(l+1)}{r^2} \right\} u_\mp = 0, \quad (3.12)$$

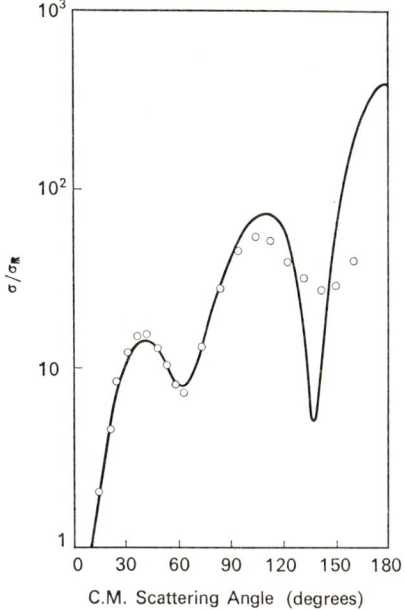

Fig. 3.2a. Proton elastic scattering fitted by the Woods–Saxon four-parameter optical model. 17 MeV protons scattered by B^{11}. $V = 50$, $W = 8$, $r_0 = 1\cdot 3$, $a = 0\cdot 5$ (adapted from Lim and McCarthy, 1966).

where $u_l^{\pm}$ are the radial partial waves for projectiles with spin up and down respectively and $U_c(r)$ is a complex central potential.

Polarization angular distributions vary with the derivative of the differential cross-section $\sigma(\theta)$. This is explained as follows. The scattering amplitude $X(\theta)$ for spin up is computed in a deeper potential than that for spin down $Y(\theta)$. They may written as

$$X(\theta) = A(\theta) + B(\theta)$$
$$Y(\theta) = A(\theta) - B(\theta). \tag{3.13}$$

The scattering is largely understood as diffraction since particles are either absorbed from the unscattered beam by the imaginary potential or refracted out of the beam direction by the real poten-

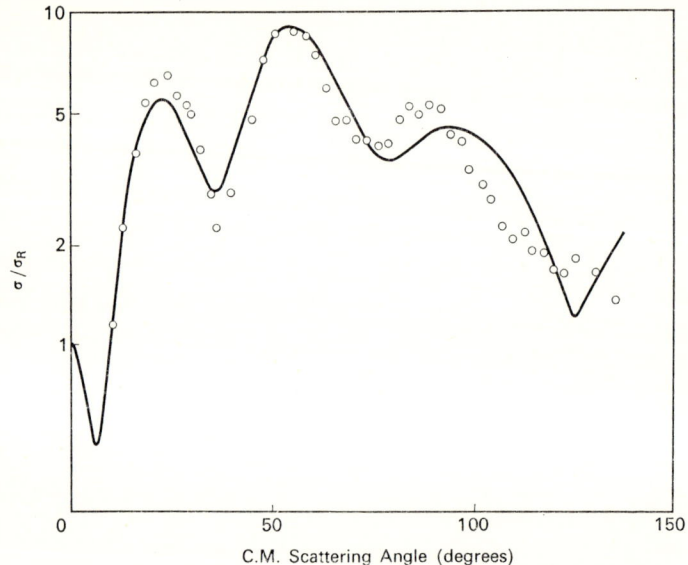

Fig. 3.2b. Proton elastic scattering fitted by the Woods–Saxon four-parameter optical model. 40 MeV protons scattered by Al^{27}. $V = 41\cdot7$, $W = 11\cdot1$, $r_0 = 1\cdot2$, $a = 0\cdot97$ (adapted from Glassgold and Kellogg, 1958).

tial. Larger potentials, both in V and R result in angular distributions whose peaks are shifted towards $\theta = 0$. In fact the peak positions depend approximately on one parameter, VR^2.

Therefore, according to these ideas, for a small angular displacement $\delta\theta$, X and Y compare with the scattering amplitude M for $V_s = 0$ as follows:

$$X(\theta) = M(\theta + \delta\theta) \simeq M(\theta) + \frac{\partial M}{\partial \theta}\,\delta\theta,$$

$$Y(\theta) = M(\theta - \delta\theta) \simeq M(\theta) - \frac{\partial M}{\partial \theta}\,\delta\theta. \qquad (3.14)$$

The polarization $P(\theta)$ is defined by

$$P(\theta) = \frac{|X(\theta)|^2 - |Y(\theta)|^2}{|X(\theta)|^2 + |Y(\theta)|^2}$$

$$\simeq \delta\theta \frac{\partial \sigma(\theta)}{\partial \theta} \sigma(\theta). \tag{3.15}$$

The inclusion of the spin–orbit coupling term remedies the fits to differential cross-sections at backward angles.

A refinement in the potential is the inclusion of a surface-peaked imaginary potential. One reason given for this is that the Pauli exclusion principle tends to inhibit reactions in the interior because many possible final states are occupied, while it is less effective in the surface. This reasoning depends on a very crude picture of a nucleus which considers various regions as having the properties of bulk nuclear matter of the local density. We will discuss bulk nuclear matter in the next chapter. For deuterons there is a special reason for surface absorption. This is the Oppenheimer–Phillips process by which the deuteron may easily break up as soon as the nuclear force begins to act. Since the binding energy of the deuteron is only 2·224 MeV a nuclear perturbation of this order of magnitude will cause the proton and neutron to separate so that the system goes into one of many possible inelastic channels. The stripping channels, in which either the neutron or proton remains bound to the target nucleus, are discussed in section 5.1.

Modern optical models include a potential like (3.10), with possibly different shape parameters r_0 and a for the real and imaginary parts, a spin–orbit potential (3.11) which is real because an imaginary spin–orbit potential gives negative partial reaction cross-sections for large enough l, and a surface absorption term

$$U_D(r) = 4ia_I W_D \frac{d}{dr} f(r, r_{0I}, a_I). \tag{3.16}$$

The Coulomb potential is of course added where necessary. It is sufficient to assume a uniform charge distribution. It is possible to determine optimum values for all the parameters phenomenologically.

In determining phenomenological potentials all the entrance channel data must, if possible, be taken into account at a particular energy. These are the differential cross-section, the angular distribution of polarization (for particles with spin), and the total reaction cross-section. For many nuclei at certain energies complete sets of such data are available for nucleon scattering. Another application of the optical model is in computing transmission coefficients for the calculation of compound nucleus reactions by the Hauser–Feshbach method. Elastic scattering at energies of a few MeV is fitted better by adding a compound elastic term including transmission coefficients calculated with the same optical model parameters as the shape elastic scattering.

The optical model may be taken seriously if the parameters are consistent from one nucleus to another at a particular energy and vary smoothly with energy, or if the differences can be understood. We expect the real potentials to behave smoothly with A, but not so much the imaginary potentials which describe the possibility of reactions and depend, especially at low energy, on the details of the nuclear structure. Some final states are easily excited, others are not. W should increase with increasing energy since more exit channels become open. We will give examples of potentials for certain probes.

PROTONS (PEREY, 1963)

$$r_0 = r_{0s} = 1\cdot 25 \text{ fm}$$
$$a = a_s = 0\cdot 65 \text{ fm}$$
$$r_{0I} = 1\cdot 25 \text{ fm}$$
$$a_I = 0\cdot 7 \text{ fm}$$
$$V_S = 7\cdot 5 \text{ MeV } (E < 17 \text{ MeV}), 8\cdot 5 \text{ MeV } (E > 17 \text{ MeV})$$
$$W = 0.$$

The dependence of V between $9\cdot 4$ MeV and $22\cdot 2$ MeV was found to be given by the following expression:

THE OPTICAL MODEL 65

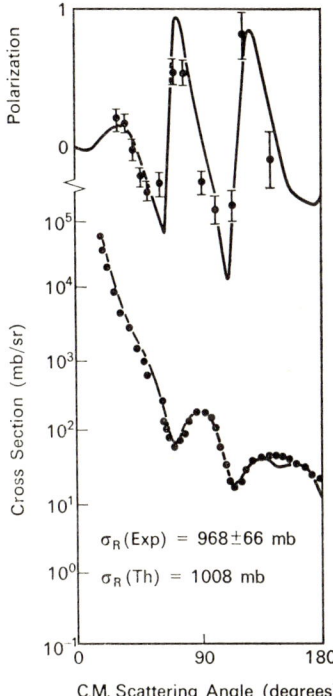

Fig. 3.3. Differential cross-section, polarization and total reaction cross-section for 17 MeV protons scattered by Cu. The calculations used the potential of Perey (1963) (adapted from Perey, 1963).

$$V = \left[53\cdot 3 - 0\cdot 55E + \left(0\cdot 4\,\frac{Z}{A^{1/3}} + 27\,\frac{N-Z}{A}\right)\right] \text{MeV}.$$

The last term reflects some dependence on nuclear structure.

The imaginary potential in this investigation was of the form (3.16). No systematic variation of W_D with E or A was found. It fluctuated about 15 MeV by several MeV.

This potential fitted all entrance channel data. Typical fits are shown in Fig. 3.3. Later work indicates that r_{0s} should be greater than r_0.

NEUTRONS (BJORKLUND AND FERNBACH, 1958)

This investigation used a gaussian surface absorption term with width parameter b. The variation of V with energy for 4·1 MeV, 7 MeV and 14 MeV was found to be similar to that for protons. In this case total cross-sections were fitted, but polarization data were not available. At 14 MeV the parameters were

$$r_0 = r_{0s} = 1\cdot25 \text{ fm}$$
$$a = 0\cdot65 \text{ fm}$$
$$r_{0I} = 1\cdot25 \text{ fm}$$
$$b = 0\cdot98 \text{ fm}$$
$$V_S = 8\cdot3 \text{ fm}$$
$$W_D = 11 \text{ MeV}$$
$$W = 0$$
$$V = 44 \text{ MeV}.$$

DEUTERONS (LEE et al., 1964)

For deuterons the potentials which fit entrance channel data are ambiguous. Entrance channel phenomena (ignoring polarization for simplicity) are described completely by the set of complex reflection amplitudes $\eta_l = \exp(2i\delta_l)$, so that potentials which produce a set of phase shifts $\delta_l + n\pi$ are all equivalent. For nucleons it is not possible to find equivalent potentials but for heavier particles it is. In the case of deuterons it is possible to resolve the ambiguity by considering non-elastic reactions. This will be discussed in Chapter V. The potential which is nearest to twice the nucleon optical model potential is the one selected. For 11 MeV deuterons scattered by Ca^{40} this is

THE OPTICAL MODEL 67

$r_0 = 0 \cdot 966$ fm

$a = 0 \cdot 846$ fm

$r_{0I} = 1 \cdot 48$ fm

$a_I = 0 \cdot 492$ fm

$V = 120 \cdot 7$ MeV

$W = 0$

$W_D = 16 \cdot 4$ MeV.

The imaginary potential extends to very large radii because the deuteron has a large radius and is so weakly bound that it can be stripped as soon as either the proton or neutron feels the nuclear force. Since the deuteron has spin 1, polarization measurements are complicated and a full set has not yet been obtained. Reaction cross-sections are not included in the present fits.

HEAVIER IONS

Optical model fits for α-particles and heavier ions are again ambiguous. They are characterized by strong absorption. The black nucleus continuum model, perhaps modified by a Coulomb potential, works quite well. For 40 MeV α-particles the positions of the first five or six maxima are fitted quite well by equation (2.62). The strong absorption is due to the fact that heavy ions almost certainly break up, become excited, or excite the nucleus.

THE FOUR-PARAMETER MODEL FOR NUCLEONS

In applications of the nucleon optical model to reactions where polarizations are not measured, and for experiments at extremely high energies, it is often convenient for computation to use the four parameter model. The approximate energy variation of V and W for C^{12} is shown in Fig. 3.4. Here we have used $r_0 = 1 \cdot 25$ fm. The value of a is approximately $0 \cdot 65$ MeV for most nuclei.

68 NUCLEAR REACTIONS

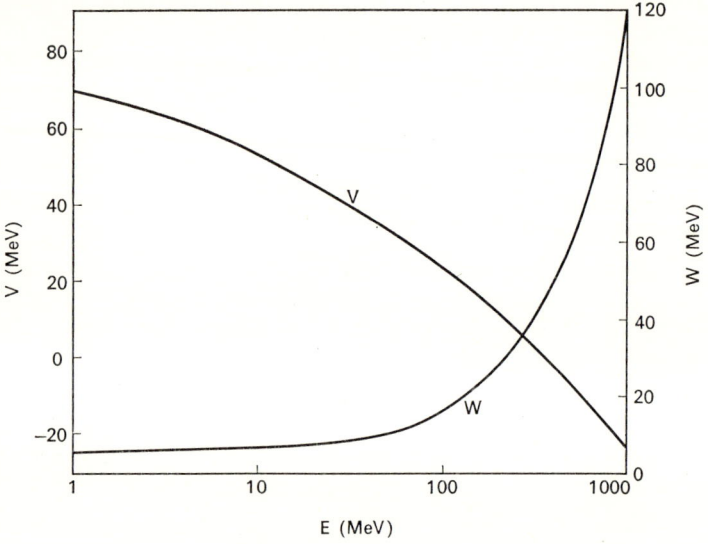

Fig. 3.4. The approximate variation with energy of the parameters V and W of the four-parameter optical model for C^{12} using $r_0 = 1\cdot 2$, $a = 0\cdot 5$.

THE NON-LOCAL MODEL FOR NUCLEONS

The complicated interaction of a nucleon with other nucleons in the nucleus is described surprisingly well by a potential operator $U(r)$ which is a function only of the coordinates of the projectile (apart from spin–orbit coupling). However, one must expect the motion of the projectile to be affected by the dynamical variables at least of neighbouring nucleons. The simplest way of including this in the one-body description is to use the following generalization of the Schrödinger equation

$$(K-E)\,\psi(r) = -\int d^3r'\, V(r, r')\psi(r'), \qquad (3.17)$$

where K is the kinetic energy operator. The function $V(r, r')$ is called a non-local potential. We may expand (3.17) in a Taylor series about $r' = r$, where $p = -i\hbar \nabla$, as follows:

$$\int d^3r' V(r, r')\psi(r') = \left[\int d^3r' V(r, r') e^{i(r'-r)\cdot p/\hbar}\right]\psi(r)$$
$$\equiv V(r, p)\psi(r). \qquad (3.18)$$

Thus a non-local potential is equivalent to an energy-dependent local potential $V(r, p)$.

It is certain that much of the energy dependence of the optical model potential can be explained in this way, although the basic energy-dependence of the nucleon–nucleon force (Brink, 1965) must also have an effect. For example the real potential, like the nucleon–nucleon potential, is repulsive at 1000 MeV (Palevsky et al., 1967).

Calculations of nucleon elastic scattering with this model were first performed by Wyatt, Wills and Green (1960). To facilitate numerical calculations the following parametrization is used.

$$V(r, r') = U\left(\frac{|r + r'|}{2}\right) H(|r - r'|). \qquad (3.19)$$

The first factor is parametrized like the local model. The second factor is taken as a gaussian of width β, the non-locality parameter.

In an extensive investigation of neutron scattering between 7 MeV and 24 MeV Perey and Buck (1962) found an energy-independent parameter set which fitted all available entrance channel data very well. The non-locality parameter β was 0·85 fm.

3.3 Particle Propagation in the Optical Model

The optical model is actually a complete description of the entrance channel aspects of a nuclear reaction. The wave function describes the motion of the projectile both inside and outside the interaction region before it is removed from the entrance channel. The description obtained by fitting entrance channel phenomena is not necessarily correct in detail. In order to fit such data it is necessary only to predict the amplitudes η_l correctly. Different potentials, particularly ones with different types of non-locality, may predict the same asymptotic wave functions but different internal wave functions. A mechanistic description of a non-elastic reaction (see Chapter V) involves an optical model description of the system while it is in the entrance channel and another one while it is in the exit channel. The transition between these channels

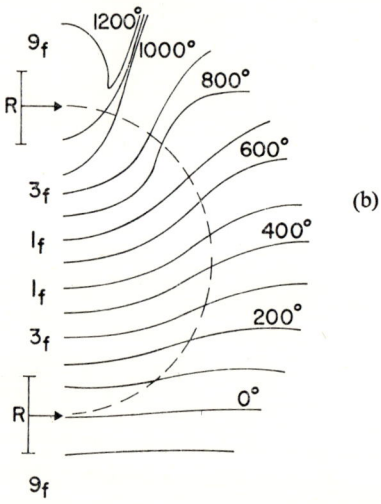

FIG. 3.5. (a) (*See insert*) The magnitude and (b) (*above*) the phase of the optical model wave function for 24 MeV neutrons scattered by Sn, calculated with a typical set of parameters (reproduced by permission, Amos, 1966).

is described by a model which includes details of nuclear structure. Prediction of non-elastic reactions provides a test of optical model wave functions inside the nucleus.

It is interesting to examine optical model wave functions in detail to see what sort of description they give of the motion of the projectile in the interior region. The magnitude and phase of the wave function for the scattering of 24 MeV neutrons on tin are illustrated in Fig. 3.5 (a) and (b). Since the scattering is axially symmetric it is sufficient to show the $\phi = 0$ plane.

FIG. 3.5 (a) (*See legend*, p. 70).

THE OPTICAL MODEL   71

The wave function is similar to a plane wave on the accelerator side of the nucleus, although some standing wave effects are caused by reflection from the front surface. Inside the nucleus the wave function is like a plane wave in a medium with a complex index of refraction. Its wavelength is shorter, so the lines of equal phase are more closely spaced. The magnitude decreases as the beam penetrates the nucleus. The most striking feature is the large peak in magnitude near the back of the nucleus. The phase diagram shows that this is a focusing effect due to the fact that the lines of equal phase inside and outside must join up.

An important fact about the phase is that for a particular value of r the projection on the scattering axis of the perpendicular distance between the lines of equal phase is roughly independent of θ, except in the focal region. The phase of the wave function at r_0 is therefore given quite closely by $K_0 \cdot r_0$, where K_0 is a propagation vector in the direction $\hat{k}$ whose value is given by the value of the complex potential at r_0. We may write $K_0 = (\beta_0 - i\gamma_0)k$.

It is interesting next to examine the probability flux

$$j(r) = \frac{\hbar}{2i\mu} (\psi^* \nabla \psi - \psi \nabla \psi^*) \quad (3.20)$$

and its divergence $\nabla \cdot j$. The flux divergence can be shown by substituting for ∇^2 from the Schrödinger equation to be

$$\nabla \cdot j = (2/\hbar)W(r)\rho(r), \quad (3.21)$$

where $W(r)$ is the imaginary part of the potential and $\rho(r) = \psi^*\psi$ is the local probability density.

The probability flux gives the quantum description of the particle flux averaged over directions. The divergence of the flux gives the probability of a reaction occurring at a particular point. If we follow Bohr's description and consider the reaction as initiated by a collision with a nucleon, the divergence of the flux tells us the probability of the collision occurring at a particular point.

In Fig. 3.6 the flux of 18 MeV α-particles in Ar^{40} is indicated by arrows and the divergence of the flux by contour lines. At the

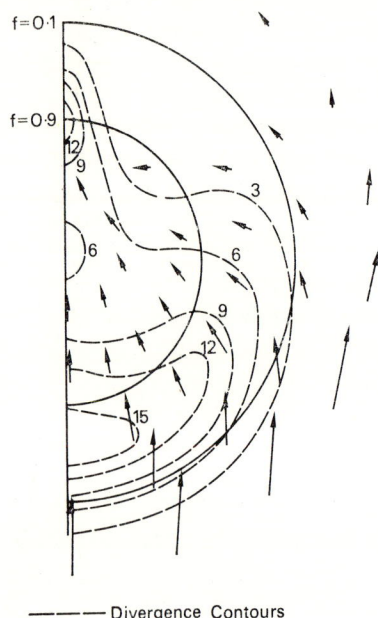

––––– Divergence Contours

Fig. 3.6. The flux of 18 MeV α-particles in Ar calculated with a typical set of parameters (reproduced by permission, McCarthy, 1959).

nuclear surface there is a clear division between particles which are repelled by the Coulomb force and ones which are refracted into the interior. Most of those refracted into the interior are absorbed, indicating that the black nucleus model should work well. The initial collision is seen to occur principally in the front surface and in the focal region where many trajectories lead to the same region of space although the probability of a particle on a particular trajectory reaching the focal region without being absorbed is small. A thorough investigation of these effects including comparison with classical trajectories was carried out by Eisberg, McCarthy and Spurrier (1959) and McCarthy (1959).

Protons and neutrons are not strongly absorbed. Their focal intensities can be up to ten times that of the external plane wave. Up to 15 per cent of all nucleon reactions are initiated in the focal region. For α-particles the absorption is so much stronger that only about 1 per cent of reactions start in the focus. The bright spot due to the focus upsets the black nucleus model for nucleon elastic scattering but is not strong enough to have a qualitative effect for α-particles.

The fact that reactions are initiated mainly in the front surface or the focal region and that the phase of the wave function is given to a good approximation for $r = r_0$ in terms of the local (complex) wave number led McCarthy and Pursey (1961) to postulate the simple five-parameter form

$$\psi(r) \simeq e^{i(\beta - i\gamma)k \cdot r} [1 + ae^{i\phi}\delta(\hat{k}r_0 - r)] \qquad (3.22)$$

as an optical model wave function to be used in a description of reactions. This form is quite close to correct in the region where reactions occur ($r \simeq r_0$), but not asymptotically, so that it does not describe elastic scattering. The second term in (3.22) represents the focus with a complex amplitude.

In describing the exit channel of a reaction the time-reversed wave function is used. We will discuss time reversal for a real Hamiltonian H. Spin-dependent Hamiltonians are complex, so the discussion applies to the simple case of spin-independent particles. Let $\psi(t)$ be a wave function satisfying the Schrödinger equation:

$$H\psi(t) = i\hbar \frac{\partial}{\partial t} \psi(t). \qquad (3.23)$$

If H is real the complex conjugate Schrödinger equation is

$$H^*\psi^* = H\psi^* = -i\hbar \frac{\partial}{\partial t} \psi^* = i\hbar \frac{\partial}{\partial(-t)} \psi^*. \qquad (3.24)$$

Thus $\psi^*(t)$ satisfies (3.23) with time reversed. The time-reversal operator is that of complex conjugation for spin-independent Hamiltonians. The time-reversed optical model wave function is

74 NUCLEAR REACTIONS

computed from a Schrödinger equation with a positive imaginary part of the potential indicating that the particle is created in the exit channel. The divergence of the flux indicates the position in space where it is created. Exit channel wave functions are represented by figures like 3.5 and 3.6 with all arrows reversed.

Consider a reaction with a particular entrance channel and exit channel. The probability amplitude of a reaction starting at r is given by the entrance channel optical model wave function. The relative probability amplitude of the particular exit channel being excited at r' is given by the exit channel optical model wave function. According to the theory of probability amplitudes the probability amplitude of both events is given by the product of the wave functions. For a complete mechanistic description of the probability amplitude we must also multiply by the amplitude for a particle absorbed at r to be emitted at r'. This amplitude will be furnished by a model for the reaction that includes a description of the nuclear structure. Since all points r and r' may be involved in the process we must integrate over r and r'.

The probability amplitude description of a quantum process is a very simple one. It is easily related to the Schrödinger equation by considering the example of elastic scattering for which the Schrödinger equation is written

$$(K - E)\psi = - V\psi. \tag{3.25}$$

The boundary condition that the asymptotic wave function must obey Huygens' principle (1.11) is built into the equation by converting it to an integral equation. This method is explained in detail in elementary quantum mechanics texts, for example Schiff (1955). We do not need the details to understand the principles. We multiply (3.25) by the inverse of the differential operator $(K - E)$, which is an integral operator $(K - E)^{-1}$ called a Green's function operator and denoted by G_0. It is chosen so that ψ has outgoing spherical waves asymptotically. The solution ψ of (3.25) must be a plane wave ϕ for $V = 0$, therefore it is written in operator notation as

$$\psi = \phi - G_0 V\psi, \tag{3.26}$$

or, explicitly,

$$\psi(r) = e^{ik \cdot r} - \frac{\mu}{2\pi\hbar^2} \int d^3r' G_0(r, r') V(r') \psi(r'). \tag{3.27}$$

The function

$$G_0(r, r') = \frac{e^{ik|r-r'|}}{|r - r'|} \tag{3.28}$$

clearly gives the correct boundary condition for large r and (3.27) is understood again in terms of Huygens' principle. The probability amplitude of finding the particle at r is the sum for all r' of the product of the probability amplitudes for the particle being at r' ($\psi(r')$) *and* interacting with the potential ($V(r')$) *and* propagating to r by means of the spherical wavelet $G_0(r, r')$. To this must be added the probability amplitude that it does not interact at all ($e^{ik \cdot r}$).

Consideration of the asymptotic form of the second term of (3.27) shows that it is $f(\theta) \exp(ikr)/r$, where the scattering amplitude $f(\theta)$ is

$$\begin{aligned}
f(\theta) &= \frac{\mu}{2\pi\hbar^2} \int d^3r' \, e^{-ik' \cdot r'} V(r') \psi(k, r) \\
&\equiv \langle \phi(k') | V | \psi(k) \rangle \\
&\equiv \langle \phi(k') | t | \phi(k) \rangle.
\end{aligned} \tag{3.29}$$

In the operator notation for scattering amplitudes used in the last two lines of (3.27) we drop the factor $\mu/2\pi\hbar^2$ for simplicity of notation. The outgoing wave vector is k'.

It is convenient to define the scattering operator or t-matrix t by (3.29) or, explicitly,

$$t\phi = V\psi = \int d^3r' \, t(r, r') \phi(r'). \tag{3.30}$$

Equation (3.29) for the scattering amplitude may be expanded in a way that gives a diagrammatic representation of the probability amplitude formulation that was first introduced by Feynman (1949) for relativistic quantum mechanics. If $V(r)$ is small we may use the plane wave $\phi(r)$ as a zeroth order approximation to $\psi(r)$. Substituting this into (3.27) we have an integral expression for $\psi(r)$ called the Born approximation. If we use this expression as an improved approximation and substitute it in (3.27) we have the second Born approximation. By iterating the integral equation in this way we develop a series expansion of $\psi(r)$, the Born series.

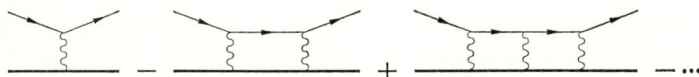

Fig. 3.7. Diagrammatic representation of the first three terms in the Born series for the elastic scattering amplitude.

The Born series for the scattering amplitude (3.29) is written in operator notation (where G_0 is an integral operator) as

$$f(\theta) = \langle \phi(k') | V - VG_0V + VG_0VG_0V - \ldots | \phi(k) \rangle. \tag{3.31}$$

The operator in this expression is the Born expansion of the t-matrix. Each term is represented by a diagram. The diagrams representing (3.31) are shown in Fig. 3.7.

Each external line represents a plane wave describing the probability amplitude of the initial or final state. Each internal line represents a Green's function $G_0(r, r')$, the probability amplitude of a particle propagating freely from r' to r. Each vertex is joined to a line representing the propagation of another particle (in this case the fixed centre of mass whose propagation is trivial and is represented by a thick line) by a wavy line representing the potential $V(r)$ which is the probability amplitude of a particle being deflected at r. Integrations are performed over the coordinates of all vertices, since the elementary scatterings can occur at any set of points in space.

IV

Nuclear Structure and Nuclear Forces

An unsatisfactory feature of the foregoing treatment of nuclear reactions is the fact that nowhere has the reaction problem been related to the basic two-nucleon interaction except in the treatment of the mean free path for 90 MeV neutrons. The object of nuclear physics is, after all, to describe nuclear phenomena in terms of the two-nucleon interaction. Historically it is true that it was quite difficult to characterize and describe reactions even qualitatively until the necessary experiments were performed and that the resonance description is basically very far removed from a two-nucleon description. In fact the development of an adequate description of the basic interaction is in itself a long and interesting story which has been told by Brink (1965) in an earlier book in this series. It was not until the late 1950's that a phenomenological description of nuclear forces over the energy range relevant to nuclear physics was achieved, owing once again to the difficulty of obtaining experimental data. This energy range extends at present to about 1000 MeV.

The optical model has however provided a simple, consistent and accurate description of the interaction of nucleons with nuclei, at least in the entrance channel. This description should be to some extent amenable to derivation from nucleon–nucleon data.

We have seen that a description of nuclear structure should enable us to predict simple non-elastic reactions. Nuclear forces will play a part in this description first through a description of the primary nucleon–nucleon collision and second through an understanding of the wave functions of the bound nuclei that is obtainable from them. The problem of understanding the states of nuclear systems in terms of nuclear forces is often called Nuclear Structure Physics. It includes unbound states or reactions, although it is simplest for the low-lying bound states. Its application to reactions is very indirect and all the more interesting for this reason.

For physicists whose main interest is the understanding of bound states the motivation for the related study of bound states and reactions is often reversed. We have already noted that nuclei are created in excited states through reactions (or occasionally radioactive decay, which we are not treating). Further, an understanding of a non-elastic reaction leads to an understanding of the initial and final bound states. Even imperfectly understood reactions are used as probes for certain properties of nuclear states such as their spins and parities.

The basic problem of nuclear structure consists of setting up a description of the Hamiltonian matrix of the system, which can be diagonalized to obtain the eigenvalues and eigenfunctions. The Hamiltonian matrix is divided into disjoint submatrices, one for each set of good quantum numbers, spin and parity, and often subdivided still further according to approximately good quantum numbers such as isospin (Brink, 1965). The numerical diagonalizing of a matrix representation of the Hamiltonian for a finite nucleus is only practicable for low-lying states. However, bulk properties, such as the optical model potentials, can be calculated from an idealized infinite system called Nuclear Matter, which has the properties of the interior of a large nucleus. In this system the wave functions are plane waves, since there is no physically distinguished reference point from which to measure positions of particles.

4.1 Nuclear Matter

Experimentally observable properties of the matter in the interior of nuclei are its density, which corresponds to a mean particle separation of about 2 fm, and the binding energy per particle, -15 MeV, which comes from a study of the masses of nuclei using the semi-empirical mass formula of Weizsäcker (1935). The object of the study of nuclear matter is to calculate these properties from nuclear forces. The Coulomb potential between protons at 2 fm is 0·7 MeV. It is neglected and protons and neutrons are treated on an equal basis.

In order to formulate a description in terms of the wave functions of particles one must use a representation of the Hamiltonian in terms of independent particle wave functions, assuming that particles move without effectively interacting. The simplest model problem which can be solved is the degenerate Fermi gas model which represents nuclear matter as a system of four different types of non-interacting particles, protons and neutrons with spin up and spin down, which obey the Pauli exclusion principle. No two particles of one type can be in the same quantum state, and all the quantum states up to a certain level k_F are filled.

At first sight it seems that such a description is so oversimplified as to be useless, particularly in view of the strong coupling ideas of Bohr. In fact the model is self-consistent up to a point and provides a very good first approximation and the key to understanding nuclear matter. The qualitative argument is due largely to Weisskopf. The first quantitative treatment of nuclear matter was due to Brueckner and collaborators. It has been reviewed and developed by Bethe (1956), Moszkowski and Scott (1961), and Gomes, Walecka and Weisskopf (1958).

The interaction of two nucleons results in each of them being scattered into final states characterized by wave vectors k_1', k_2' which are different from their initial wave vectors k_1, k_2. Unless the magnitudes k_1, k_2 are greater than k_F, this interaction is forbidden by the Pauli principle. The probability amplitude for the interaction is, in the centre of mass system of particles 1 and 2,

$$K(k', k) = \int d^3r \int d^3r' \, e^{-ik' \cdot r'} \, K(r, r') \, e^{ik \cdot r}$$
$$= \int d^3r' \, e^{-ik' \cdot r'} \, v(r') \psi(k, r'), \qquad (4.1)$$

where k, k' are the relative wave vectors and r, r' are the relative position vectors. The non-local function $K(r, r')$ is the K-matrix of Brueckner, which is the t-matrix or scattering operator for the collision of two nucleons in nuclear matter with the Pauli principle restriction included. Such collisions are not observable because nuclear matter cannot spontaneously increase its energy. The amplitudes (4.1) are virtual scattering amplitudes. We will see what part they play in the solution of the two-body Schrödinger equation.

With the inclusion of an operator Q_F which imposes the Pauli principle condition, the two-body wave equation is the Brueckner–Bethe–Goldstone equation:

$$\left(\frac{\hbar^2}{2\mu}\nabla^2 + \epsilon\right)\psi(r) = Q_F v(r)\psi(r). \tag{4.2a}$$

The wave function $\psi(r)$ has the boundary condition that it is a plane wave asymptotically, since real scattering cannot occur. Phase shifts are zero.

The right-hand side of the free Schrödinger equation (with $Q_F = 1$) may be expanded in terms of a complete set of plane wave intermediate states with relative wave numbers k'.

$$v(r)\psi(r) = (2\pi)^{-3}\int d^3r'\ v(r')\psi(r')\int d^3k'\ e^{ik'\cdot(r-r')}. \tag{4.2b}$$

The operator Q_F restricts k' to values for which both k_1' and k_2' are greater than k_F.

$$\begin{aligned}Q_F v(r)\psi(r) &= (2\pi)^{-3}\int d^3r'\ v(r')\psi(r)\int_{k_F}^{\infty} d^3k'\ e^{ik'\cdot(r-r')}\\ &= (2\pi)^{-3}\int_{k_F}^{\infty} d^3k'\ e^{ik'\cdot r}\int d^3r'\ e^{ik'\cdot r'}\ v(r')\ \psi(k, r')\\ &= (2\pi)^{-3}\int_{k_F}^{\infty} d^3k'\ e^{ik'\cdot r}\ K(k', k). \end{aligned} \tag{4.3}$$

The expectation value of the total energy of nuclear matter in its ground state is

$$E_0 = \langle \Psi_0 | H | \Psi_0 \rangle \tag{4.4}$$

where Ψ_0 is the wave function of the ground state. We make the approximation that only pair interactions are important so that the pair wave functions ψ represent the ground state. They are calculated by solving the same equation as for two-body collisions in free space except that the boundary condition is different and that the interaction term differs from the one in free space by

$$(2\pi)^{-3} \int_0^{k_F} d^3k'\, e^{ik'\cdot r}\, K(k', k). \tag{4.5}$$

In this theory matrix elements of the potential are not calculated. Realistic local nuclear forces have infinite repulsive cores at about 0·4 fm, so that their matrix elements are infinite. Virtual scattering amplitudes (4.1) are however finite because ψ is zero where v is infinite. The term (4.5), when calculated with realistic nuclear forces, does not strongly affect the wave function and it reaches its asymptotic form, which is the same as the form for non-interacting particles, at distances of the order of k_F^{-1} as we will now show.

A physical argument for the resemblance of nuclear matter wave functions to Fermi gas wave functions is that we expect $\psi(r)$ to differ from the non-interacting wave functions $\phi(r)$ only because of the interaction term (4.3) which depends on virtual transitions to states outside the Fermi sphere. These transitions affect the short-range part of the wave function, $r < k_F^{-1}$, which corresponds to close collisions and therefore to momentum transfers greater than k_F. In fact for very small r, the term (4.5) is negligible and $\psi(r)$ is the same as the wave function for two nucleons interacting in free space.

It is possible, using realistic nuclear forces, to calculate the correct density and binding energy per particle. This depends on the short-range or strong correlations. However we are more interested in the fact that the independent particle model of nuclear matter is justified, given the density. It is justified if the healing distance of about k_F^{-1}, at which the wave functions resemble Fermi gas wave functions, is less than the mean spacing of nucleons.

Using the standard statistical mechanics of a Fermi gas and the

82 NUCLEAR REACTIONS

value $r_0 = 1\cdot 1$ fm, which is obtained from high-energy electron scattering experiments that probe the charge distribution of nuclei very accurately, we find

$$k_F = (3\pi^2\rho/2)^{1/3} = (9\pi^2/8\pi)^{1/3}/1\cdot 1 \text{ fm}^{-1}$$
$$= 1\cdot 4 \text{ fm}^{-1}. \tag{4.6}$$

This corresponds to a Fermi energy of about 40 MeV and a healing distance of about 1 fm, which is less than the mean spacing.

The Fermi gas model is therefore very largely self-consistent. Only for states very near the top of the Fermi energy distribution will the virtual transition amplitudes be important. We say that the uppermost particles in the energy distribution have weak or long-range correlations while the bulk of the nucleons act in many applications as if they are independent. This fact is central to the understanding of nuclear physics in terms of nuclear forces. Once it is accepted we can derive many simple results which are close to quantitatively correct.

For example we have seen how Fernbach, Serber and Taylor were able essentially to calculate the imaginary part of the optical model potential at high energies by assuming free collision cross-sections for two nucleons modified by a factor 3/5 which comes from a rough geometrical calculation for determining which collisions result in states outside the Fermi sphere.

The real potential acting on a nucleon is obtained by knowing the average separation energy of the topmost nucleon from mass measurements. It is about 8 MeV. The depth of the potential is therefore about 50 MeV, made up of the kinetic energy and the separation energy. This checks quite closely the figures found for the optical model at low energies. At higher energies the wavelength is much shorter and incident nucleons begin to see individual nucleons of the target. The optical model potential, like the two-nucleon potential, becomes more repulsive at high energies and in fact is repulsive at 1000 MeV. At high energies the imaginary part increases sharply as meson production channels open (see Fig. 3.4).

More detailed calculations of optical model potentials at high energies have been performed by Watson (1958) and by Kerman, McManus and Thaler (1959), but the results are only semiquantitatively correct.

4.2 The Shell Model

The independent particle model for finite nuclei is the shell model. It is assumed that the motion of each particle is independent of the motion of the others except that it is attracted by a one-body potential, the shell model potential, caused by the collective effect of all the others, and that it obeys the Pauli exclusion principle. States of orbital angular momentum l have magnetic degeneracy $2l + 1$. Single particle wave functions for bound states in a spin-independent potential are eigenstates of l^2, l_z and the radial quantum number n which is the number of nodes in the radial wave function, not counting the nodes at zero and infinity. They are represented as

$$\psi_l^m(\mathbf{r}) = r^{-1} u_{nl}(r) Y_l^m(\theta, \phi), \tag{4.7}$$

where $u_{nl}(r)$ is the solution of the radial Schrödinger equation (1.10) with negative energy. The boundary condition is that $u_{nl}(r)$ tends to zero for large r. In the exterior region it is called a spherical Hankel function $h_l^{(1)}(i\beta r)$ for uncharged particles. It is the corresponding Coulomb function for charged particles. The exterior wave function is given by the binding energy E.

$$\beta^2 = -2\mu E/\hbar^2. \tag{4.8}$$

Energy levels are named by the number $n + 1$ followed by a letter of the old spectroscopic notation (s, p, d, f, g, h, i, ...) denoting l. Levels with larger n and l have higher energies. The nucleons of each type are assumed to fill up the levels in the shell model potential, with $2(2l + 1)$ in each. There will be quite large energy gaps between certain groups of levels (or between each level for very light nuclei). Such clusters of levels are called shells. Since real and virtual transitions to different many-body states (for example ones with the spins of nucleons in the topmost level aligned differently) are easy for unfilled shells and difficult for filled shells, nuclei will exhibit periodic properties with increasing A. Closed shell nuclei for example have anomalously small level densities and anomalously large binding energies.

The anomalous (magic) numbers of protons or neutrons were tabulated by Elsasser (1934) in the light of binding energy information. Numbers with less-pronounced anomalies are given in parentheses. The numbers are

$$2, (6), 8, (14), 20, 28, 50, 82, 126.$$

The shell model in a central harmonic oscillator potential was discussed in detail by Bethe and Bacher (1936). It predicted the lower magic numbers, 2 for the filled $1s$ shell, 8 for filled $1s$ and $1p$ shells where the level spacings are large, but failed to predict which level spacings were small and which were large for higher levels, so that the wrong magic numbers were obtained. The model fell into disfavour as a simple description of nuclear properties, although it was stressed by Bethe and Bacher that it has a more fundamental use as a complete set of basis wave functions in terms of which the actual nuclear wave function can be expanded. Such an expansion will have a manageably small number of terms if the basis states are reasonably realistic approximations to the eigenstates, that is if the representation of the Hamiltonian is approximately diagonal. Shell model wave functions at least represent one major correlation between nucleons, their space localization.

In 1948 it was discovered independently by Mayer (1948) and by Haxel, Jensen and Suess (1949) that the addition of a spin–orbit coupling term to the shell model potential would split each l subshell (except for $l = 0$) into two, for total angular momentum $j = l \pm \frac{1}{2}$, in such a way that the level clustering would explain the correct magic numbers. The levels are illustrated in Fig. 4.1. The number of protons or neutrons in each level is now $(2j + 1)$. Values of the potential parameters are comparable with those of the low-energy optical model. Not only are the magic numbers explained, but the lowest energy levels of nuclei consisting of a

NUCLEAR STRUCTURE AND NUCLEAR FORCES 85

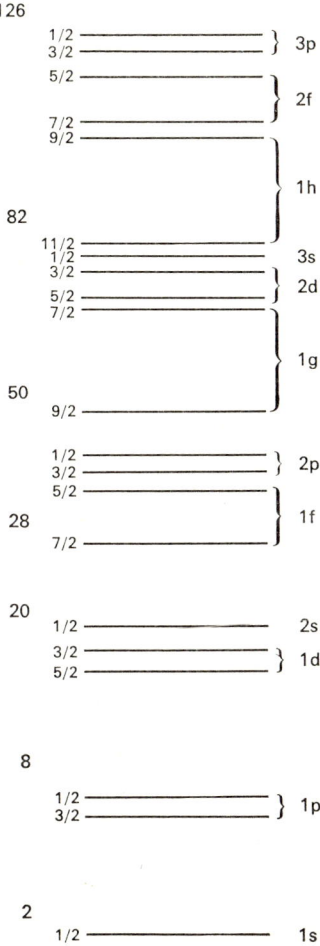

FIG. 4.1. Single-particle energy levels computed in a finite square well potential with spin–orbit coupling (adapted from Haxel, Jensen and Suess, 1950). The spin of each level is indicated on the left and the cumulative total of particles is on the far left at each major energy gap.

closed shell plus one nucleon have the spin and parity of the single particle states immediately above the closed shell. The parity is the parity of l. This is a property of the spherical harmonics. Single particle states with spin–orbit coupling are denoted with the j-value as a subscript. For example the $1p$ level splits into the $1p_{3/2}$ and $1p_{1/2}$ levels, the one with the higher j-value being lower in energy. A hole in a closed shell behaves like a particle with the spin and parity of the shell.

The shell model explains another very general fact about nuclei. Even nuclei, that is ones with even numbers of protons and neutrons, have spin zero in the ground state. The topmost nucleons tend to form pairs with opposite spins because if two similar nucleons had the same spin one would have to be in a single particle state of higher energy. The pairing tendency is reinforced by the attractive pairing forces between nucleons which lower the total potential energy in paired states.

The independent particle shell model does not predict the total binding energies of nuclei at all well, just as the Fermi gas model of course predicts unbound nuclear matter instead of a binding energy of 15 MeV per particle. However, the inclusion of nuclear forces corrects this. Nuclear forces were first included in the shell model by Heisenberg (1935), who used harmonic oscillator wave functions and a spin-dependent gaussian model for the two-body force. With this simple model the problem is one of coupled oscillators, which can be transformed by the normal coordinate transformation and solved exactly. Qualitatively reasonable results were obtained.

Modern nuclear structure theory, made possible by high-speed computers, uses independent particle wave functions, which are antisymmetrized products of A single-particle wave functions, as a basis for a matrix representation of the Hamiltonian with realistic two-body forces. Sometimes both particles and holes are included in the basis and assumed to interact. The matrix elements are the virtual transition amplitudes, just as in the theory of nuclear matter, and the representation is truncated by omitting single-particle states whose energy is so high that the virtual transition amplitudes are small. In this way the spectroscopic properties of many of the lower excited states can be predicted. Nuclear eigenstates are linear combinations of independent particle wave functions with coefficients given by the virtual transition amplitudes.

For the ground state it is possible to solve the shell model problem with interactions self-consistently. This means that the density reproduced by the wave functions from a variational calculation using a shell model basis is used to determine a potential which gives an independent particle basis for a second calculation, and the process is continued until it converges. This is called a Hartree–Fock calculation. It is found that lower binding energies are obtained for some nuclei if degrees of freedom are included which represent non-spherical ground states. We may say that for these nuclei, which are far removed from closed shells, the long-range correlations take the form of ground state deformations. The lower energy levels are explained as collective rotations of the deformed nucleus.

Since the shell model representation of a nucleus is merely a convenient mathematical device, it is necessary only to consider a large enough basis to describe any nuclear state. However, nuclei in between closed shells require a very large and complicated basis with many matrix elements contributing to the calculation. It is more convenient to describe them in terms of collective degrees of freedom. For further reading on the shell model the reader is referred to the review article by Elliott and Lane (1957) and the book by Mayer and Jensen (1955).

4.3 Collective Models

The strong-coupling model of nuclear structure is the liquid drop model of N. Bohr (1936). The unification of this model with the particle concept of nuclei has been achieved in a series of papers by A. Bohr and Mottelson. A paper which gives a good introduction to the ideas and contains earlier references is Bohr and Mottelson (1957). The motivation for the development of collective models was the discovery in spectroscopic experiments of certain regularities in the properties of the lower excited states of heavier nuclei. We will explain some of these regularities in the course of explaining the models.

A collective description of nuclear states requires fewer coordinates than a particle description. Collective coordinates describe changes in shape or orientation of the nuclear density distribution while the volume is constant. Nuclei which are spherical in their ground state may have low excited states described by vibrational degrees of freedom. Nuclei with deformed ground states require less energy for rotation than vibration, so the lower states are rotational.

We will denote the collective coordinates by α. The normal coordinates of collective surface oscillations are the expansion parameters of a function $R(\theta', \phi')$ in terms of spherical harmonics. This function represents the shape of the surface, which is assumed to be sharp, not diffuse.

$$R(\theta', \phi') = R_0 [1 + \Sigma_{\lambda\mu} a_{\lambda\mu} Y_\lambda^\mu(\theta', \phi')]. \tag{4.9}$$

The coordinates θ', ϕ' are measured with respect to axes fixed in the body. The collective effects are produced by the average motion of particles. It is assumed that the frequencies associated with particle motion are much higher than those associated with collective motion so that the many-body wave function $\psi_{n\nu}(x)$ may be factored into two parts. $\phi_n(x, \alpha)$ describes the motion of particles, whose space, spin and isospin coordinates are denoted by x, in a nucleus whose collective coordinates have the set of values α. $\psi_\nu(\alpha)$ is the probability amplitude of the particular set α occurring.

$$\Psi_{n\nu}(x) = \psi_\nu(\alpha)\phi_n(x, \alpha). \tag{4.10}$$

This is the adiabatic approximation. $\phi_n(x, \alpha)$ may be given for example in terms of shell model wave functions. The sets of quantum numbers n, ν characterize different particle and collective states.

Surface vibrations are considered to be harmonic to a first approximation. A quantum of vibrational energy in which the

$Y_\lambda^\mu(\theta', \phi')$ mode is excited is called a phonon of type $\lambda\mu$. The eigenstates are degenerate in μ, so that we have a $(2\lambda + 1)$-dimensional oscillator. Harmonic oscillators have equally spaced energy levels. The number of phonons of type $\lambda\mu$ is denoted by $n_{\lambda\mu}$. The number of phonons of type λ is given by

$$N_\lambda = \Sigma_\mu n_{\lambda\mu} = 0, 1, 2, \ldots.$$

The theory of the quantum harmonic oscillator shows that the angular momentum of a phonon of type λ is $\lambda\hbar$.

The ground state has no phonons. For an even nucleus it is a 0+ state. Nuclear states are denoted by a number specifying the total spin J followed by a sign indicating the parity.

Vibrations with $\lambda = 0$ correspond to a change of volume of the nucleus. This requires too much energy for such a state to be among the lowest. For $\lambda = 1$ the centre of mass oscillates, which is unphysical. The lowest-order vibration we must consider is $\lambda = 2$. In fact one of the regularities noted among large even nuclei is that the first excited state has spin and parity 2+. In a spherical nucleus this is a one-phonon quadrupole vibration. Two-phonon quadrupole states may be either 0+, 2+, or 4+, corresponding to the ways of adding two angular momentum vectors of length 2. In the simple collective model these states are degenerate. The model does not predict in what way the degeneracy is resolved by the nuclear forces, and in fact vibrational nuclei are characterized by a group of states 0+, 2+ and 4+ in haphazard order with varying A. The vibrational energy is of the order of 1 MeV. The model predicts that the ratio of the excitation energy of the states of the 0+, 2+, 4+ triplet to that of the first 2+ state is 2. In many nuclei this prediction is fulfilled within a factor of about 30 per cent. The harmonic approximation for the vibrations is not very good. In some nuclei there is a low-lying 3− state, which is described as a one-phonon octupole vibration.

A deformed nucleus is considered to a first approximation to be an ellipsoid of revolution or a symmetric top. The quantum mechanics of a symmetric top is given in some elementary text books, for example Pauling and Wilson (1935). The angular momentum about the z' axis (the symmetry axis fixed in the body) is $I_{z'}$. The angular momentum about the space-fixed z axis is I_z. The total angular momentum is I.

The eigenstates of the symmetric top are the rotation functions D^I_{MK}. They are simultaneous eigenstates of $I_{z'}$, I_z and I^2.

$$I_{z'} D^I_{MK} = K D^I_{MK},$$
$$I_z D^I_{MK} = M D^I_{MK}, \qquad (4.11)$$
$$I^2 D^I_{MK} = I(I+1) D^I_{MK},$$

where $|K| \leqslant I$ and $|M| \leqslant I$.

The rotation functions are the coefficients in the transformation of the spherical harmonic $Y^\mu_\lambda(\theta', \phi')$ from the body-fixed system to the space-fixed system which is obtained by a rotation through the Euler angles α, β, γ.

$$Y^\mu_\lambda(\theta, \phi) = \Sigma_\rho Y^\rho_\lambda(\theta', \phi') D^\lambda_{\mu\rho}(\alpha, \beta, \gamma). \qquad (4.12)$$

For rotational states we will again consider only the simplest case as an illustration. This is an even nucleus which is not rotating about the z' axis so that $K = 0$. The lowest states correspond to rotations in the space-fixed system about the x or y axes for which the effective moment of inertia is $\mathscr{I}$. They all have even parity. The energy levels are

$$E(I) = E_0 + \frac{\hbar^2}{2\mathscr{I}} I(I+1), \qquad (4.13)$$

where E_0 is a constant.

The lowest states are the ground state rotational band

$$0+,\ 2+,\ 4+,\ \ldots$$

The theoretical ratios of energy transitions E_I from the ground state to the state I are

$$E_4/E_2 = 10/3,$$
$$E_6/E_2 = 7,$$
$$E_8/E_2 = 12.$$

In the rotational regions of the periodic table, $A \simeq 24$, $150 < A < 190$ and $A > 230$, these ratios are observed within a few per cent. The effective moment of inertia increases with increasing mass, but is much less than the value calculated by assuming that the nucleus is a rigid body. This is expected from the fact that for heavy nuclei not all the nucleons take part in the rotational correlations.

The collective models of nuclei are related to our understanding of reactions in terms of two-nucleon forces much less directly than is the shell model. In order to describe a reaction, we must know the probability amplitudes of the target and residual nuclei being in the relevant states. That is we must know their wave functions. With the shell model the nuclear wave function is a relatively simple linear combination of independent particle wave functions with coefficients given by the virtual two-body transition amplitudes.

States of some nuclei which may be considered as vibrational have been calculated in the shell model (Goswami and Pal, 1963; Gillet and Vinh Mau, 1964). A very complicated linear combination involving a large number of wave functions of the independent particle–hole basis is used. In this way we can describe the same states with two models, showing in principle how the models are related.

The main parameter of the symmetric quadrupole rotational model is the deformation parameter β_2. This is used to specify the ground state shape function $R(\theta', \phi')$.

$$R(\theta', \phi') = R_0 [1 + \beta_2 Y_2^0(\theta', 0)]. \tag{4.14}$$

It may be calculated from nuclear forces by using the Hartree–Fock method (see for example Pal and Stamp, 1967). Transition rates both for γ-decay and nuclear reactions depend on β_2, so its value provides a bridge between reaction calculations and calculations involving nuclear forces. In the following chapter we will discuss the use of nuclear models in reaction calculations.

For further reading on collective models the reader is referred to the review article by Moszkowski (1957) and references therein.

V

Direct Interactions

IN 1947 Serber proposed an extension of Bohr's picture of the collision of a particle with a nucleus, which was suggested by the results of high-energy collision experiments at Berkeley. The incident particle first collides with a nucleon or cluster of nucleons in the target and shares its energy, but it is quite likely that one of the two particles, or even both, have enough energy left to emerge from the nucleus without a further collision, because the mean free path for collisions is of the order of the nuclear radius. Such a reaction is called a direct interaction. If a second collision does occur it is then assumed that Bohr's mechanism is dominant. The products of the second collision do not escape immediately.

Direct interactions occur most probably for reactions in which the nucleus is left in one of its lower-lying excited states. The mechanism was first suspected in the case of (d, n) reactions by Lawrence, Livingstone and Lewis (1933) and calculated by Oppenheimer and Phillips (1935), even earlier than Bohr's statement of the strong coupling philosophy. Yields of fast neutrons in reactions initiated by 200 MeV deuterons from the Berkeley synchrocyclotron in 1947 were far too large to be accounted for by the evaporation theory. Evidence was soon forthcoming that the angular distribution for both low and high energies was concentrated in the forward direction rather than being symmetrical about 90° as predicted by the Bohr assumption. Roberts and Abelson (1947) noted forward peaking in (d, n) reactions at 15 MeV and Helmholz, McMillan and Sewell (1947) reported forward peaking at 190 MeV. The latter experiment was explained by Serber (1947a) using a semiclassical model.

Direct interactions are expected to take place in a time comparable with the time for a particle to cross the nuclear volume once. They are the analogues for non-elastic reactions of the shape elastic scattering part of the elastic scattering. The direct interaction amplitude is defined as the slowly-varying amplitude about which the energy fluctuations due to the resonance terms in the actual amplitude average to zero. They are calculated by means of potential scattering mechanisms which involve only a few degrees of freedom of the system.

5.1 Stripping

In 1950 and 1951 angular distributions for several (d, p) and (d, n) reactions with resolved final states were measured. Burrows, Gibson and Rotblat (1950) and Holt and Young (1950) reported (d, p) angular distributions for resolved final states using O^{16} and Al^{27} respectively as targets. They were characterized by a large forward peak and one or more lesser peaks, very reminiscent of a diffraction pattern. The former data were fitted by Butler (1950), using a theory which enabled the orbital angular momentum of the transferred particle in the residual nucleus to be determined. The (d, n) angular distribution for O^{16} was measured by El Bedewi, Middleton and Tai (1951) and analysed by Newns using an unpublished Born approximation theory due to Bhatia and Huang. Again the spectroscopic application was noted. We will first consider these simple theories which bring out the point of the determination of angular momenta.

The theory of stripping angular distributions was formulated by Butler (1950, 1951) in terms of elastic scattering wave functions for particles in initial and final states. Since at that time it was impossible to compute these functions, Butler replaced them by plane waves and discovered that the resulting approximation reproduced the diffraction-like angular distribution and was even capable of giving quite a detailed fit to the shape of the first peak.

The formulation in the Born approximation of Bhatia, Huang, Huby and Newns (1952) is reproduced in Part 2. This paper is

94 NUCLEAR REACTIONS

chosen because it is simpler than Butler's. In terms of our probability amplitude ideas the Born approximation is represented by the lowest-order diagram which will explain the effect. This is shown in Fig. 5.1.

The nucleus is represented by a thick line denoting an assumed fixed centre for the nucleon–nucleus potential $V(r, \sigma, \xi)$, where r and σ denote the nucleon position and spin coordinates and ξ denotes the coordinates of the target or core nucleus. The subscript p or n will be used on r and σ to denote proton or neutron

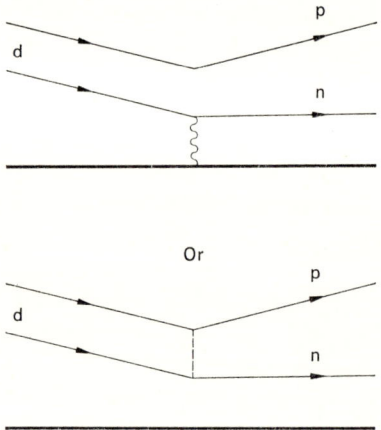

Fig. 5.1. Diagrammatic representation of the stripping amplitude in the Born approximation.

where necessary. The initial state is the direct product of internal wave functions $\chi_d(r_n - r_p, \sigma_n, \sigma_p)$ for the deuteron (double line) and $\chi_i(\xi)$ for the target (thick line) and the wave function $\exp[i k_d \cdot (r_n + r_p)/2]$ of their relative motion. The potential V is assumed not to act in the initial state. The final state is denoted by $\chi_f(r_n, \sigma_n, \xi)$ for the neutron bound to the target (thick–thin line), the internal (spin) wave function $\chi_p^*(\sigma_p)$ for the proton and the plane wave $\exp(i k_p \cdot r_p)$ for their relative motion. The probability amplitude for the neutron transfer process may be regarded as $V(r_n, \sigma_n, \xi)$,

DIRECT INTERACTIONS 95

which holds the neutron bound (wavy line), or as $v(r_n-r_p)$, the neutron–proton interaction (dotted line) which can cause break-up of the deuteron in the three-body situation where the necessary momentum is transferred to the core. If one of these interactions seems more plausible than the other to the reader, he need only consider the time-reversed situation (pick-up), which is known to be equivalent since nuclear interactions are time-reversal invariant. The roles of the two potentials are then reversed and the other potential would seem more plausible by equivalent reasoning.

The paper of Bhatia *et al.* uses $V(r_n, \sigma_n, \xi)$ for the interaction. The theory differs slightly from Butler's although both are based essentially on the Born approximation. The differences are noted in the paper. Nowadays it is more usual to use $v(r_n-r_p)$ for the interaction. We make the transformation

$$\boldsymbol{R} = (\boldsymbol{r}_n + \boldsymbol{r}_p)/2$$
$$\boldsymbol{\rho} = \boldsymbol{r}_n - \boldsymbol{r}_p. \tag{5.1}$$

The stripping amplitude is most simply understood by dropping the spin coordinates with the intention of replacing them when necessary. Since spin functions are orthogonal the spin integrations result in constant factors due to magnetic degeneracies. The amplitude is given by

$$f(\boldsymbol{k}_p, \boldsymbol{k}_d) = \int d^3 r_n \int d^3 r_p \, e^{-i\boldsymbol{k}_p \cdot \boldsymbol{r}_p} \, \psi_l^{m*}(r_n) v(\rho) \chi_d(\rho) \, e^{i\boldsymbol{k}_d \cdot \boldsymbol{R}}, \tag{5.2}$$

where $\chi_f(r_n, \sigma_n, \xi)$ is approximated by the single particle wave function $\psi_l^m(r_n)$. The approximation is now made that the proton and neutron of the deuteron are in a relative *s*-state and that

$$v(\rho)\chi_d(\rho) = V_0 \delta(r_n - r_p). \tag{5.3}$$

This is the zero range approximation. It is not very good, since the binding energy of the deuteron is 2·2 MeV, so that the tail of its wave function (the part outside the interaction region) is given by (4.8) as $r^{-1} \exp(-0\cdot 24 r)$. Here we have used the fact that $h_0^{(1)}(i\beta r) = r^{-1}\exp(-\beta r)$. The size of the deuteron is therefore characterized by $\beta r = 1$ or $r \simeq 4$ fm. Butler assumed that the reaction takes place in the surface of the nucleus outside a cut-off radius R_B, the Butler radius. In this way the absorption of particles inside the nucleus into other channels is taken into account.

The stripping amplitude (5.2) reduces to

$$f(k_p, k_d) = \int_{R_B} d^3r \, e^{i(k_d - k_p) \cdot r} \psi_l^m(r). \quad (5.4)$$

From (5.4) it is clear that the angular distribution in the model depends on l, the angular momentum transfer $Q = k_d - k_p$ and the Butler radius R_B. The differential cross-section for a zero spin target is given by squaring (5.4), performing the angular integrations, and summing over the final states which are degenerate in m. (This degeneracy could be resolved by polarization measurements, but in fact is not.)

$$d\sigma(\theta)/d\Omega \propto \left| \int_{R_B}^{\infty} r \, dr \, j_l(Qr) \, u_{nl}(r) \right|^2, \quad (5.5)$$

where $u_{nl}(r)$ is the radial shell model wave function of (4.7) and $j_l(\rho) = (\pi/2\rho)^{\frac{1}{2}} J_{l+\frac{1}{2}}(\rho)$ is a spherical Bessel function. Since $u_{nl}(r)$ falls rapidly with r for $r > R_B$, the integrand is fairly well localized to R_B. A good approximation is

$$d\sigma(\theta)/d\Omega \propto |j_l(QR_B)|^2. \quad (5.6)$$

This function has a large initial peak whose maximum is nearer to $0°$ for smaller l, larger R_B and higher energy. The peaks for the first few spherical Bessel functions are shown in Fig. 5.2.

The Butler–Born approximation was immediately recognized as a very good spectroscopic tool, since the first peak for reactions in which l was known could not be fitted with the incorrect l without using an improbable value of R_B. The values of R_B for correct l were about $2A^{1/3}$. The large radius is due mostly to the large radius of the deuteron and the ease of stripping it in the nuclear periphery.

Further consideration of the stripping mechanism shows that the probability of stripping is proportional to the probability of the nucleus accepting a single neutron, that is of the single-neutron-plus-target configuration being one of the wave functions which contributes significantly to an expansion of the actual wave function $\chi_f(r_n, \sigma_n, \xi)$ of the final nucleus in terms of a convenient basis. This probability is called the spectroscopic factor. It must be less than the number of equivalent neutrons in the final state.

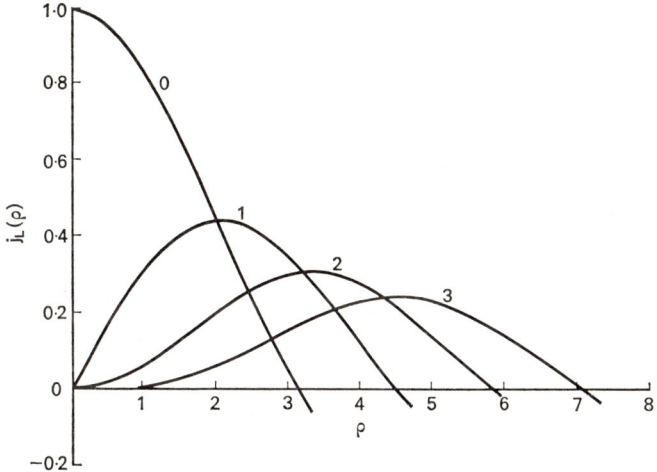

Fig. 5.2. The first peaks in the spherical Bessel functions.

In fact the simple Butler theory does not predict it within a large factor mainly because plane waves do not give a sufficiently good description of the particle interactions with the core.

In the application of stripping to nuclear structure it is often interesting to know the ratio of spectroscopic factors for two reactions that put a neutron into the same single particle state ψ_l^m which is a component of the wave functions of two different final states χ_1, χ_2. If the amplitudes with which the normalized single-particle wave function occurs in the expansions of χ_1 and χ_2 are a_1 and a_2 respectively, the ratio of spectroscopic factors is a_1^2/a_2^2. A theory of nuclear structure predicts values for a_1 and a_2

which can be tested in this way. We do not need an accurate description of the reaction. We assume that the differential cross-section has a shape depending only on l and optical model parameters of the nucleus which are nearly independent of the state. Its magnitude is assumed to be proportional to a_f^2 for final state χ_f. These assumptions have been tested many times and found adequate. Thus the stripping reaction itself is a valuable spectroscopic tool even without a detailed reaction theory. Spectroscopic information about more complicated states may be obtained by the same technique applied to two-nucleon stripping. The use of stripping in spectroscopy is discussed at length by Macfarlane and French (1960).

The simplest of all applications of the stripping reaction is in determining the energies of unoccupied single-particle states. The energy of the outgoing particle is assumed to be given by the energy of the single-particle state into which the other particle is stripped, once the masses of all the particles have been taken into account. A review of neutron levels determined by (d, p) reactions has been given by Cohen (1965).

A more detailed reaction theory which is often successful in predicting spectroscopic factors and at least the first peak in angular distributions is the distorted wave Born approximation (DWBA). This theory replaces the plane waves $\exp(i\mathbf{k}\cdot\mathbf{r})$ of (5.2), describing the probability amplitude of finding an unbound particle, by optical model wave functions $\chi^{(\pm)}(\mathbf{k}, \mathbf{r})$ for the elastic scattering of the particle at the appropriate energy. The plus or minus refer respectively to the situation with outgoing spherical waves or the time-reversed situation with ingoing spherical waves.

In diagram language the DWBA includes all the diagrams in which the incident particle interacts with the nucleus through an optical model potential V_i before the two-nucleon event and the emerging particle interacts through V_f after the two-nucleon event. The two-nucleon potential acts only once in each diagram, hence the name Born approximation. The approximation is easily handled by high-speed computers.

A typical DWBA fit is shown in Fig. 5.3. The stripping cross-section in this approximation depends on the optical model wave functions in the interaction region, not merely in the exterior region as does the elastic scattering cross-section. Hence stripping and other direct interactions may be used to resolve ambiguities in optical model potentials. In particular the stripping angular distributions for the smaller of the equivalent sets of potentials for

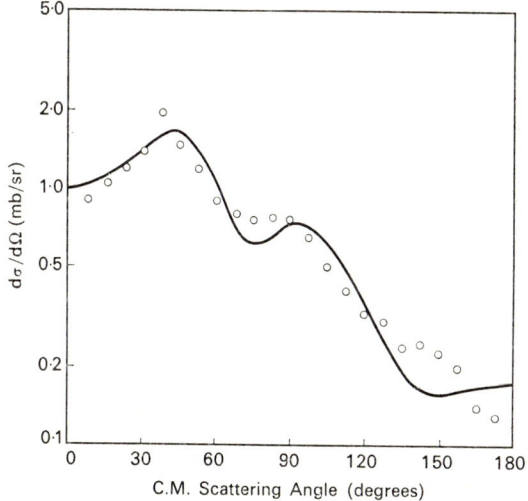

Fig. 5.3. The distorted wave Born approximation for the 7 MeV (d, p) reaction on Ca^{44}, leaving Ca^{45} in its ground state. The spectroscopic factor is 1.

deuterons are extremely bad. The best potential for stripping is the one of the equivalent sets for elastic scattering whose real part is closest to the sum of the real parts of the neutron and proton optical model potentials. The stripping angular distributions for different potentials reflect differences in the internal wave functions. The main difference is that the focus moves towards the centre of the nucleus as the real potential is increased.

The reader is referred to a paper by Lee *et al.* (1964) and a review article by Bassel (1964) for an extensive account of the use of the DWBA.

5.2 Inelastic Scattering

The direct interaction mechanism for inelastic scattering of protons and (p, n) reactions was suspected by Cohen (1953) as a result of experiments which again showed that the high-energy end of the spectrum of emitted particles was weighted much more strongly than the evaporation model would allow. Gugelot (1954) obtained experimental plots of n/ES_p, where n is the number of out-going particles with energy E and S_p is a transmission factor included to compensate equation (2.44) for the fact that particles of the Maxwell distribution may fail to penetrate to the outside. One such plot is shown in Fig. 5.4. On a logarithmic scale we expect approximately a straight line, whose slope determines the temperature, on the basis of (2.44). For higher energies the transmission factor is close to unity. As Serber expected, the straight line is observed for higher excitation energies (lower emitted particle energies), but the excitation of the lower-lying states is much more probable than the evaporation model predicts.

Eisberg and Igo (1954) found that the angular distributions for the 31 MeV (p, p') reaction on Pb were peaked forward and that the cross-section for reactions in the part of the spectrum that did not obey the Maxwell distribution law was $0 \cdot 15 \, \pi R^2$, where πR^2 is the geometrical cross-section. Simple considerations of billiard ball collisions showed that about 30 per cent of reactions would be initiated in the surface rim where the density is below its interior value (the density distribution is obtained from high-energy electron scattering). The (p, p') direct interaction mechanism would account for half of these absorption events.

The importance of the direct interaction mechanism at least for higher incident energies was thus established for head-on collisions of particles and nuclei. Because of the large separation of the proton and neutron in the deuteron, there had been justification for considering stripping to be perhaps exceptional.

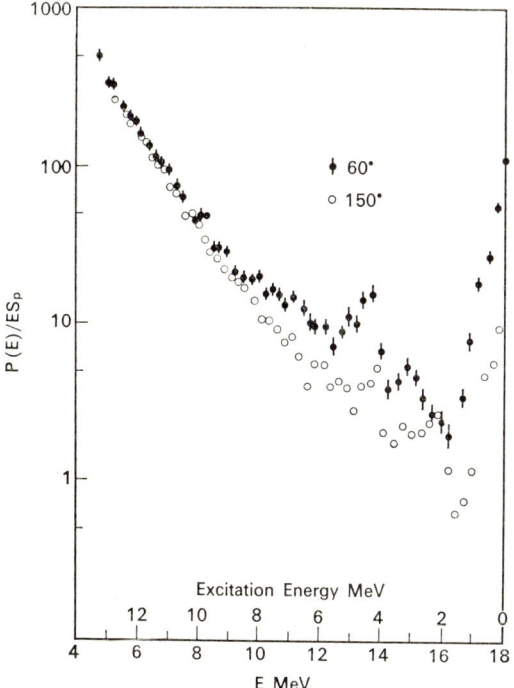

Fig. 5.4. The distribution of energies of outgoing protons for 31 MeV protons incident on Fe. The function $P(E)/ES_p$ is plotted on an arbitrary logarithmic scale (adapted from Gugelot, 1954).

The angular distributions for (p, p') and (p, n) reactions were not nearly so easy to understand as in the case of stripping. The Butler stripping formalism was extended to such reactions, for example (p, n), by Austern, Butler and McManus (1953). Essentially the incident probability amplitude of (5.2) is replaced by the probability amplitude for an ingoing proton and a bound neutron $e^{i\mathbf{k}_p \cdot \mathbf{r}_p} \psi_l^m(\mathbf{r}_n)$, where the neutron is assumed to be initially in a single particle state. The angular distribution is again given approximately by (5.6) where l this time is the difference between initial and final orbital angular momenta.

Experimental angular distributions for nucleons in fact did not have this shape at all. The differential cross-section was rather slowly-varying, and certainly not symmetrical about 90°. The shape changed considerably for energy changes of a few MeV. Examples are shown in Fig. 5.5. Levinson and Banerjee (1957) managed to obtain qualitatively correct angular distribution shapes

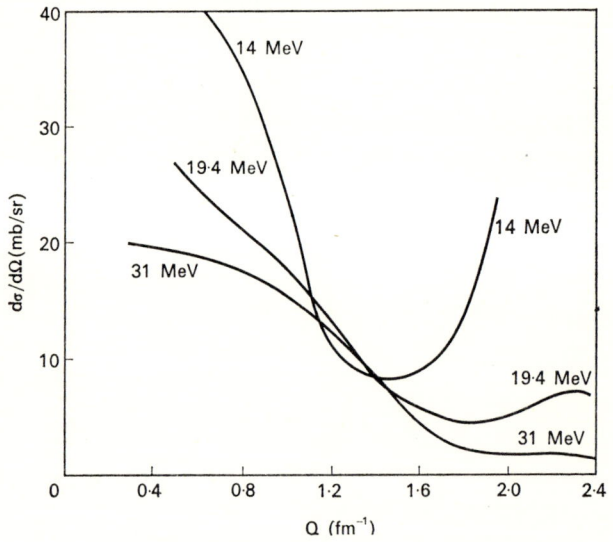

FIG. 5.5. Experimental momentum transfer distributions for the inelastic scattering of protons of different energies to the first excited (2+) state of C^{12} (adapted from Levinson and Banerjee, 1958).

for proton inelastic scattering at different energies to the first 2+ state of C^{12} by using the single-particle model for nuclear structure and describing the reaction in the distorted wave Born approximation. This was the first full DWBA calculation for a nuclear reaction. The complications of the distorted-wave calculation were such that its evident qualitative correctness did not provide much insight into the reasons for the angular distributions having the particular shapes that were observed.

At this stage in the development of the understanding of direct interactions the question of the basic mechanism of inelastic scattering came up. Are direct interactions initiated by a collision of the incident particle with a nucleon, or is a collective interaction involved which would leave the nucleus most probably in a final state described by the collective model? Two dramatic answers were provided.

It was reasoned by Eisberg that if the initial nucleon–nucleon collision mechanism were valid there should be a finite chance of both the incident and struck particles escaping from the nucleus. A collision of two free particles of equal mass in which one is at rest results in an angle of 90° between the two outgoing directions. Hence, even allowing for the fact that the struck particle should be moving with about the Fermi energy and that all trajectories would be bent to some extent by the optical potentials, there should be some vestige of a maximum at 90° in the distribution of angles between the two out-going protons in a $(p, 2p)$ experiment.

A $(p, 2p)$ experiment at 40 MeV on Ni was performed by Griffiths and Eisberg (1959) using the Minnesota linear accelerator. The experiment required a very long time for minimum counting statistics since the instantaneous beam intensity must be so low that it is certain that two protons counted in coincidence both came from the same target nucleus. The low duty cycle of the accelerator was a severe handicap. Particles knocked out of states with different binding energies were not resolved. They had been unsuccessfully looked for in a previous $(p, 2p)$ experiment by Cohen (1957). A very broad peak at 90° was just observed within the limits of experimental error and the total $(p, 2p)$ cross-section was shown by a classical collision argument to be consistent with the initial nucleon–nucleon collision mechanism. The total cross-section argument is probably more significant in the light of later knowledge of $(p, 2p)$ angular distributions (see section 5.3) although the general idea that the outgoing protons carry off most of the incident momentum is correct.

In 1959 elastic and inelastic α-particle scattering experiments at about 40 MeV were performed by McDaniels, Blair, Chen and

Farwell (1960) at the University of Washington cyclotron. The elastic angular distributions showed a series of several regularly-spaced maxima up to about 90° as predicted by the black nucleus model. The inelastic angular distributions to particular final states were rather similar. Blair realized that the inelastic angular distributions for 2+ excitations were like the ones that had been predicted for the excitation of a 2+ collective state by Drozdov (1955) and Inopin (1956) using the model of a deformed or deformable black nucleus.

The model of collective excitations of a black nucleus was examined very thoroughly by Blair (1959) with extremely good results. The first success was the positive identification of low-lying octupole (3−) excitations. Not only were the positions of the peaks explained, but a semiquantitative model for determining the collective parameter β_λ was established and the phase rule for predicting the spin and parity of single-phonon excitations was discovered. This rule states that odd parity excitations have periodic angular distributions which are in phase with those of elastic scattering and out of phase with those of even parity excitations. The rule is not obeyed by 4+ excitations which are explained as two-photon excitations. The basic ideas of the theory were reviewed and developed and the phase rule introduced in a paper given by Blair at the Kingston conference on nuclear structure in 1960. This paper is reproduced in Part 2.

Both the initial two-body collision mechanism and the excitation of collective states are therefore established. The ideas are reconciled by realizing that the initial collision may be with one of many nucleons near the top of the energy distribution. This nucleon is correlated with the others in such a way that the collision excites a collective state. Such a collective excitation is very much more probable than a single particle excitation. In fact the states that are well-described by single particle models (for example in closed-shell-plus-one nuclei) have cross-sections which are smaller by a factor of ten or more.

The black nucleus model of Blair gives a simple explanation for the shapes and magnitudes of angular distributions for inelastic scattering of strongly-absorbed particles including He^3 and deuterons, but not for protons or neutrons. The plane wave Born approximation gives similar shapes but not magnitudes. At first sight it seems strange that the less-strongly-absorbed particles have angular distributions that are much less like the plane-wave Born approximation in which the effect of the optical model potential is included only by cutting off the radial integration at R_B.

A simple explanation of angular distributions in the DWBA was given by McCarthy and Pursey (1960) using the approximate optical model wave functions (3.22). The qualitative explanation is that the absorption and focusing localize the reaction to regions of configuration space that are smaller than the nuclear volume. The uncertainty principle therefore tells us that the corresponding momentum transfer distributions will be characteristic of the localization. The peaks in the spherical Bessel function $j_l(QR)$ characterize a region of space whose size is given by R. The focal region, however, is much smaller, so that one expects interference from a momentum transfer amplitude that has a much slower angular variation. Thus it is the transparency of the nucleus, which gives a strong focus, that is responsible for the departure from the spherical Bessel function shape.

In the case of α-particles the focus is so weak that only the first term of (3.22) is important. This is a more complete description of the localization than Butler's radial cut-off, since it describes the fact that the magnitude of the optical model wave function decreases with decreasing angle from π to 0, measured from the beam direction for each optical model wave function. The angular distribution is given by a spherical Bessel function with a complex argument, since the first term of (3.22) is just a plane wave with a complex wave number. McCarthy and Pursey computed a radial integral like (5.5) with $u_{nl}(r)$ replaced by a gaussian function describing the localization of the collective interaction to the surface. One of the fits so obtained is shown in Fig. 5.6.

Kromminga and McCarthy (1962) included the focus in a calculation of (p, p') inelastic scattering from the first 2+ state of C^{12} at different energies, obtaining remarkably good shapes

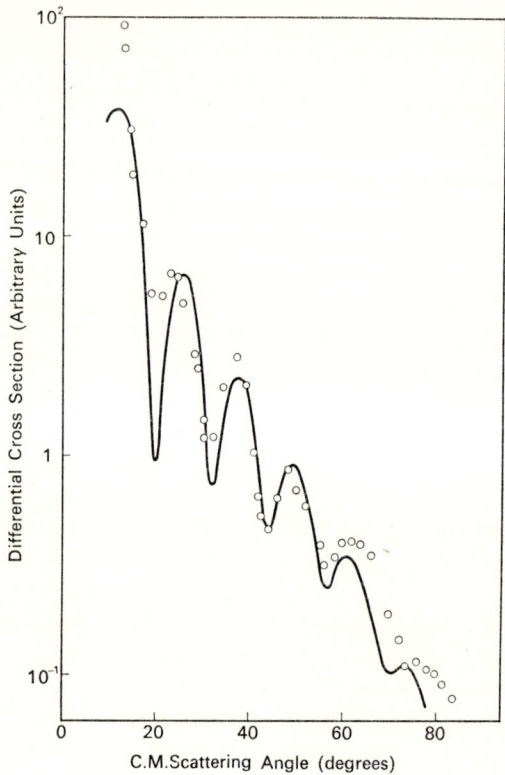

FIG. 5.6. The inelastic scattering of 41·7 MeV α-particles to the first excited (2+) state of S^{32} fitted by the DWBA model of McCarthy and Pursey (adapted from McCarthy and Pursey, 1961).

considering the simplicity of the model. The fit at 40 MeV is shown in Fig. 5.7. The essential correctness of the understanding of angular distribution shapes in terms of localization is established by this work. The success of the Butler theory for predicting the first stripping peak is due to the strong absorption of the deuteron, so that the focus is unimportant.

At this stage we will explain the DWBA in more detail. The

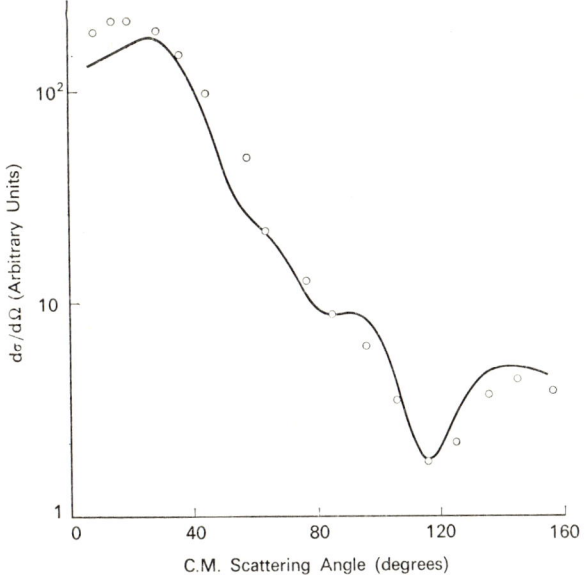

FIG. 5.7. The inelastic scattering of 40 MeV protons to the first excited (2+) state of C^{12} fitted by the DWBA model of McCarthy and Pursey (adapted from Kromminga and McCarthy, 1962).

DWBA for collective interactions differs from that for interactions involving a two-body mechanism by the use of a different expression for the probability amplitude of a particle interacting at r in the entrance channel and appearing at r' in the exit channel. For the two-body mechanism this nuclear probability amplitude is the amplitude for the particle belonging to the nucleus to be transferred from the initial bound state $\psi_{l'}^{m'}(r')$ to the final bound state $\psi_l^m(r)$ by the action of the potential $v(r, r')$. If we include isotopic spin in the description of the nucleons they are indistinguishable and the matrix element must be antisymmetrized.

$$f(k', k) = 2^{-1/2} \int d^3r \int d^3r' \, \chi^{(-)*}(k', r)\psi_l^{m*}(r')v(|\,r - r'\,|) \\ \times [\psi_{l'}^{m'}(r)\chi^{(+)}(k, r') - \psi_{l'}^{m'}(r')\chi^{(+)}(k, r)]. \tag{5.7}$$

For collective excitations inelastic scattering is closely related to elastic scattering. The projectile can be considered to bounce off the nucleus leaving it merely translating, or translating and vibrating, or translating and rotating. The nucleus is described by an optical model potential with extra degrees of freedom describing its non-spherical shape. The theory of collective excitations is called the generalized optical model. The potential is

$$\tilde{V}(r, \xi) = U(r) + V[r - R(\theta', \phi')], \qquad (5.8)$$

where $R(\theta', \phi')$ is the shape function for the nuclear surface. The second term in (5.8) is the non-spherical potential operator which is responsible for the collective excitation. It is assumed to be restricted to the nuclear surface and to depend only on the distance from the mean surface $R(\theta', \phi')$.

For a rotation of a nucleus with symmetric quadrupole deformation we transform the shape function (4.14) to the space-fixed system by means of the rotational transformation (4.12), so that the potential causing excitation is

$$v(r, \Omega) = \beta_2 \Sigma_p Y_2^{p*}(\Omega) D_{p0}^2(\hat{S}) R_0 dU/dr. \qquad (5.9)$$

This is the first order term in a Taylor expansion of (5.8) about R_0. The derivative dU/dr of the Woods–Saxon optical model potential describes the surface localization of the potential. The Euler angles $\hat{S}$ describe the orientation in space of the nuclear symmetry axis.

The probability amplitude of the nucleus changing from one of the rotational eigenstates $D_{MK}^I(\hat{S})$ to another under the action of (5.9) is given by a matrix element of three rotation functions where the integration is over $\hat{S}$. Standard theory of angular momentum reduces the matrix element for nuclear states of spin I and I' to

$$V_{II'} = \frac{\beta_2 R_0}{(4\pi)^{1/2}} (-1)^I (2I+1)^{1/2} C_{I'2I}^{000} \frac{dU}{dr} \Sigma_p Y_2^{p*}(\Omega), \qquad (5.10)$$

where $C_{I'2I}^{0\,00}$ is a Clebsch–Gordan coefficient. Note the similarity of (5.10) to a product of single-particle wave functions for angular momentum transfer $l = 2$. The angular factor is the same and the radial factor is peaked at the surface just like the product of two radial wave functions $u_{nl}(r)$ and $u_{n'l'}(r)$, where $|l-l'| = 2$.

The fact that the space localization determines the shape of the angular distribution tells us that the single-particle excitation mechanism and the collective excitation mechanism give very similar angular distribution shapes. This is confirmed by detailed calculations. There is, however, a large difference in the numerical factors which multiply the nuclear probability amplitude in each case. The correct prediction of absolute magnitudes of cross-sections is necessary to confirm the validity of a reaction theory.

The nuclear probability amplitude for the excitation of vibrational states is obtained from the quantum theory of harmonic vibrations. We will not consider the details. A brief description is given in Blair's article in Part 2. For $l = 2$ it is the same as (5.10). In this case β_2 describes the mean square deformation.

The DWBA with complete partial wave expansions of the optical model wave functions was first applied to collective excitations by Rost and Austern (1960). Values of β_2 were predicted which checked very well with the values obtained from electromagnetic transitions.

In the DWBA the probability of exciting channels other than the one under study is taken into account by the imaginary part of the optical model potential. For scattering of protons from C^{12}, for example, the cross-section for inelastic scattering from the first $2+$ state is a very large proportion of the entire reaction cross-section. Clearly the optical model wave functions used in the DWBA for the first $2+$ state should not include the possibility of exciting this state, which is being treated explicitly. Tamura and Terasawa (1962) noticed that the cross-section for the $2+$ excitation varied with energy in such a way that it appeared to be coupled to elastic scattering. The elastic cross-section was large when the $2+$ cross-section was small and vice-versa. For such reactions it is necessary to solve equations which couple the channels.

We will consider that the nucleus is described by an optical model potential $\tilde{V}(r, \xi)$ which already has an imaginary part representing the excitation of the channels which are not to be coupled explicitly. The projectile coordinates are r and the collective coordinates of the target are ξ. The Schrödinger equation for the entire scattering process is

$$[K(r) + H_0(\xi) + \tilde{V}(r, \xi) - E]\Psi(r, \xi) = 0. \qquad (5.11)$$

The Schrödinger equation of the target in an eigenstate α is

$$[H_0(\xi) - E_\alpha]\psi_\alpha = 0. \qquad (5.12)$$

The set of quantum numbers of the target state is denoted by α. The total wave function $\Psi(r, \xi)$ is expanded in the complete set of target states $\psi_\alpha(\xi)$. We will truncate the expansion to include only the states to be coupled and assume that the others make a small contribution to Ψ which is described by the imaginary part of V. Note that this is done when we write the optical model equation for just the ground state ψ_0. We are here generalizing the optical model to include the lower excited states $\alpha = 0, \ldots, N$.

$$\Psi(r, \xi) = \Sigma_{\alpha=0}^{N} u_\alpha^{(+)}(r)\psi_\alpha(\xi). \qquad (5.13)$$

We obtain a set of coupled differential equations for the expansion coefficients $u_\alpha^{(+)}$ by substituting (5.13) in (5.11), multiplying on the left by $\psi_\beta^*(\xi)$ and integrating over ξ. The equations are simplified by using (5.12) and the orthonormality of the $\psi_\alpha(\xi)$. There is one equation for each β.

$$[K(r) - E + E_\beta]u_\beta^{(+)}(r) = -\Sigma_\alpha \int d\xi\, \psi_\beta^*(\xi)\tilde{V}(r, \xi)\psi_\alpha(\xi)u_\alpha^{(+)}(r). \qquad (5.14)$$

Using the separation (5.8) of $\tilde{V}$ into a spherical term U and nonspherical term V and describing the matrix element of V for states ψ_β, ψ_α as $V_{\beta\alpha}$, we have the following coupled equations:

$$[K(r) + U_{\beta\beta}(r) - E + E_\beta]\, u_\beta^{(+)}(r)$$
$$= -\Sigma_{\alpha \neq \beta}\, V_{\beta\alpha}u_\alpha^{(+)}(r), \beta = 0, \ldots, N. \qquad (5.15)$$

The spherical potentials $U_{\beta\beta}(r)$ are assumed to be the same for all the low-lying states β. The model is the generalized optical model. If only $\beta = 0$ is included in the set of target states, (5.15) is the optical model equation and $u_0^{(+)}(r)$ is the optical model wave function with outgoing spherical wave boundary conditions.

Consider the 2+ state, $\beta = 2$.

$$[K(r) + U_{22}(r) - E + E_2]u_2^{(+)}(r) = - V_{20}u_0^{(+)}(r). \quad (5.16)$$

We may uncouple the equations by using for $u_0^{(+)}(r)$ the solution of the ground state $\beta = 0$ equation with V_{02} neglected. This is the DWBA.

The generalized optical model was first computed in a simplified way using plane waves by Chase, Wilets and Edmonds (1958). Coupled channel calculations for 0+ and 2+ states were performed by Buck (1963). These and later calculations by Tamura (1965), which couple more channels, constitute a great triumph for the generalized optical model. Shapes and magnitudes of many states are fitted and the collective excitation model confirmed. The variation with A of Γ_0/D is also predicted much more accurately than with the single-channel optical model (see Fig. 3.1).

The single-particle collision mechanism is more difficult to calculate. Modern calculations for nucleon reactions use a modification of (5.7), or a linear combination of such matrix elements if the nuclear wave function is expanded in shell model wave functions. The modification replaces the potential v by the two-nucleon t-matrix t. Thus we include all diagrams for scattering of the two nucleons. The approximation is the distorted wave t-matrix approximation (DWTA).

At high energy ($E > 150$ MeV) a great simplification is possible. We consider the matrix element in momentum space. The distorted waves vary rapidly with momentum while the two-body t-matrix does not. It is taken outside the integration. On squaring the matrix element we have the zero range DWBA cross-section multiplied by the two-nucleon cross-section at the appropriate energy, which is known. This is the impulse approximation. It is explained in terms of plane waves in section 5.3. It has been successfully

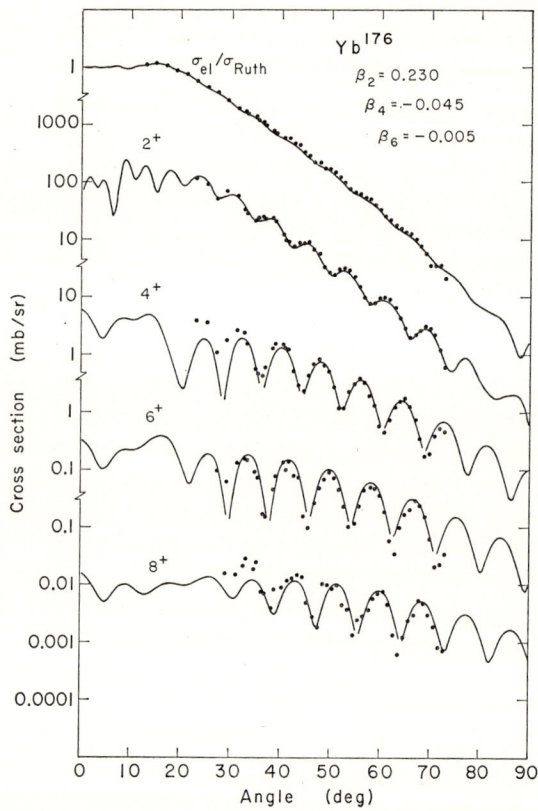

FIG. 5.8. The scattering of 50 MeV α-particles from the first five (rotational) states of Yb^{176} fitted by the generalized optical model with one set of optical model parameters. Deformation parameters are adjusted to fit the magnitude of the cross-section for each state. (Reproduced by permission of D. L. Hendrie.)

applied by Haybron and McManus (1965) to predict magnitudes and shapes of angular distributions for states which are described by a linear combination of single-particle wave functions. In the impulse approximation the antisymmetry between the two nucleons is taken into account in the two-nucleon cross-section.

At lower energy the impulse approximation cannot be made. We need a representation for the t-matrix. Calculations by Amos, Madsen and McCarthy (1967) use a local, spin-dependent function chosen to reproduce a convenient subset of p-p scattering data. The fully antisymmetric matrix element (5.7) is computed. Results are qualitatively correct for absolute magnitudes, but the exact angular distribution shape is not yet understood.

One of the states for which proton inelastic scattering was successfully calculated by Haybron and McManus is the first 2+ state of C^{12}. This state exhibits the large inelastic scattering cross-section characteristic of collective states. In fact a collective model calculation by Haybron (1966) fitted the data. However, the complicated shell model description which corresponds to the collective description is known in this case and was used in an impulse approximation calculation. This calculation is a microscopic description in quantum mechanics of our physical picture of the direct interaction. The excitation of the 2+ state is described by particle collisions. However, the nuclear probability amplitude is described by wave functions which express the correlations between the particles in the nucleus.

5.3 The $(p, 2p)$ Reaction

The $(p, 2p)$ reaction is the most direct way of observing single protons in a nucleus. The first experiments were performed by Chamberlain and Segrè (1952) at 340 MeV. Although different states were not resolved it was shown that the distribution of momentum transfer was roughly consistent with that expected from the collision of the incident proton with protons moving with the momentum distribution expected on the basis of the shell model. The momentum distribution of a particle in a particular single-particle state is the square of the Fourier transform of its space wave function. Thus it was established that two-body collisions without further significant interaction of either particle occur in nuclei. We have seen that more quantitative confirmation of the importance of the single collision mechanism in reactions, even at much lower energy (40 MeV), was obtained by Griffiths and Eisberg (1959).

The importance of the $(p, 2p)$ reaction for understanding nuclear structure became evident when it was possible to resolve the energies of groups of particles in the final state. The separation energy of a proton is assumed to be the binding energy of the single-particle level. For the topmost level the reaction is complementary to the (d, n) reaction which puts a proton into an unoccupied single particle level. However, the $(p, 2p)$ reaction probes the occupied single-particle levels by putting a proton hole into one of them.

The first $(p, 2p)$ experiment in which separation energies corresponding to single-particle levels were resolved was performed by Tyrén, Hillman and Maris (1958) on several light nuclei, of which C^{12} is an example, using the Uppsala synchrocyclotron with 180 MeV proton beam energy. For C^{12} a sharp peak at 16 MeV separation energy indicated the $1p_{3/2}$ level and a peak at 36 MeV with a width of 10 MeV indicated the $1s$ level.

The widths of the $(p, 2p)$ peaks are due to the fact that a hole is created in the internal nuclear matter. In contrast to the case of particles in nuclear matter, the Pauli principle and conservation of energy do not prevent real transitions for holes. A hole with negative energy acts like a particle with positive energy. It feels a complex potential and has a complex eigenvalue whose imaginary part is the width of the level. Low-lying holes can make more transitions than ones nearer the top of the energy distribution. Therefore, the $1s$ states in large nuclei have the largest widths. Holes in the topmost state cannot make transitions. The width of the $(p, 2p)$ cross-section curve for such a reaction is zero. Correct widths for measured hole states were calculated by Köhler (1966) using an extension of the theory of nuclear matter. He predicted widths of 20 MeV or more for low-lying single hole states in heavy nuclei. Such a width would prevent the state from being identified experimentally. This is the factor limiting the usefulness of the $(p, 2p)$ reaction as a probe for single particle states. For example the $1s$ state is observed in O^{16} but not in Ca^{40}.

The $(p, 2p)$ differential cross-section depends on both outgoing proton angles and one outgoing proton energy (for a definite single-particle state). Its distribution is measured in various ways, the most common being to keep the counters at equal angles θ on opposite sides of the incident direction and to consider coplanar events with almost equal energies. The distribution of the angle θ is plotted. According to the semiclassical argument of the previous section, $\theta = 45°$ if the struck particle is stationary, and the spread of the distribution indicates the spread of the distribution of momenta of the struck particle. This picture omits possible refraction of particles by the nuclear potential and the possible excitation of other channels. Its quantum analogue is the plane wave Born approximation. In a review paper by McCarthy, reproduced in Part 2, it is shown that the plane wave scattering amplitude is proportional to

$$T = \left[\int d^3r \; e^{i(k_0 - k_L) \cdot r} \, v(r) \right] \left[\int d^3r' \; e^{i(k_0 - k_L - k_R) \cdot r'} \, \psi_l^m(r') \right].$$
(5.17)

The subscripts 0, L, R indicate the incident, left and right beams respectively, $v(r)$ is the two-nucleon potential and $\psi_l^m(r)$ is the single-particle wave function. The exponent in the second factor is the momentum transfer, which is equal in this model to the momentum transfer of the struck particle. The second factor is the momentum space wave function of the struck particle and its square is the momentum distribution. The first factor, when squared, gives the p-p cross-section, which is known at different energies. The energy at which it should be taken is not well determined by the model, but this does not matter at higher energies where it varies slowly. This approximation is the impulse approximation.

Equation (5.17) leads for $l > 0$ to a distribution of θ which has two peaks corresponding to the first peak of a spherical Bessel function starting at about 45° where the momentum transfer is minimum. The peak for smaller θ corresponds to negative momentum transfer, that is the case where the struck particle moves in the same direction as the incident particle. The peak for larger θ corresponds to head-on collisions. The symmetric geometry of the experiment restricts the considerations to collinear motion in the plane wave model. For $l = 0$ there is one peak because $j_0(Qr)$ is maximum for $Q = 0$.

These peaks are characteristic of $(p, 2p)$ angular distributions at high energy where the energy is so much greater than the optical model potentials that the plane wave approximation is qualitatively adequate. The first measurement of an angular distribution with resolved separation energy and coplanar symmetric geometry was due to Tyrén, Hillman and Maris (1960). The characteristic peaks were observed. Momentum transfer distributions for the p-state proton of C^{12} had been independently measured at Harwell by Gooding and Pugh (1960). They observed the distribution of the outgoing energy of each proton at different angles for an incident energy of 153 MeV.

Since $(p, 2p)$ angular distributions at high energy depend strongly on the single-particle wave functions they can be used to obtain a value of at least one parameter describing the extension in space of the single-particle state. The tail of the radial wave function is already known because the separation energy is measured. The r.m.s. radius can be determined. This was done for several light nuclei by Lim and McCarthy (1964) who fitted data obtained at the Orsay synchrocyclotron at 155 MeV (Riou, 1965). The values of proton r.m.s. radii for $1s$ and $1p$ states checked the value for the summed charge distribution obtained by high-energy electron scattering (Hofstadter, 1956). This calculation used the DWBA and showed that the r.m.s. radius is determined uniquely by fitting the overall width of the angular distribution. The overall width is practically independent of the optical model parameters. Earlier calculations using modified plane waves were not sufficiently detailed to establish the uniqueness of the radius determination.

The determination of the shapes of single-particle wave functions by means of the $(p, 2p)$ reaction confirms a qualitative prediction of self-consistent shell model calculations with realistic nuclear forces (Brueckner, Lockett and Rotenberg, 1961). The single-particle potential well is narrower and deeper for deeper single-particle states.

For a more detailed review of some of the earlier $(p, 2p)$ work the reader is referred to an article by Jacob and Maris (1966). Early experiments at 150 MeV or higher energies were unable to resolve particular states of the residual nucleus. Analysis of these experiments assumed that the residual nucleus was left in its ground state.

The first $(p, 2p)$ experiment in which particular states of the residual nucleus were identified was performed with a C^{12} target by Pugh, Hendrie, Chabre and Boschitz (1965) using the 50 MeV proton beam from the 88-in. azimuthally-varying-field cyclotron at the Lawrence Radiation Laboratory, Berkeley. The ground state of B^{11} was excited more strongly than the higher states by a factor of about ten. The ground state angular distribution was analysed in the DWTA, which was discussed in the previous section on inelastic scattering. The DWTA gives the correct answer, not the Born approximation answer in the absence of the B^{11} core. It is semi-quantitatively successful up to the present time (Pugh *et al.*, 1967), but shows that a greater understanding of reactions involving three bodies is required.

In addition to information about nuclear structure, the $(p, 2p)$ experiment also provides a test of models for the interaction of two protons in the presence of other nuclear matter. In the DWTA at 50 MeV it has so far been found that the model which best fits p-p scattering in free space provides the best fit to the $(p, 2p)$ reaction.

The $(p, 2p)$ reaction is not so far well enough understood to be a good tool for the spectroscopy of the residual nucleus. Its use in this respect has been discussed by Jain and Jackson (1967).

Bibliography

AMALDI, E., D'AGOSTINO, O., FERMI, E., PONTECORVO, B., RASETTI, F. and SEGRÈ, E., *Proc. Roy. Soc. (London)* **A522**, 149 (1935).
AMOS, K. A., *Nucl. Phys.* **77**, 225 (1966).
AMOS, K. A., MADSEN, V. A. and MCCARTHY, I. E., *Nucl. Phys.* **A94**, 103 (1967).
ASTON, F. W., *Proc. Roy. Soc. (London)* **A115**, 487 (1927).
AUSTERN, N., BUTLER, S. T. and MCMANUS, H., *Phys. Rev.* **92**, 350 (1953).
BARSCHALL, H. H., *Phys. Rev.* **86**, 431 (1952).
BASSEL, R. H., Some applications of the distorted wave approximation for direct nuclear reactions, *Few Nucleon Problems*, Vol. II, p. 199, edited by M. Cerineo, Federal Nuclear Energy Commission of Yugoslavia, Zagreb, 1964.
BETHE, H. A., *Rev. Mod. Phys.* **9**, 69 (1937).
BETHE, H. A., *Phys. Rev.* **57**, 1125 (1940).
BETHE, H. A., *Phys. Rev.* **103**, 1353 (1956).
BETHE, H. A. and BACHER, R. F., *Rev. Mod. Phys.* **8**, 82 (1936).
BETHE, H. A. and PLACZEK, G., *Phys. Rev.* **51**, 450 (1937).
BHATIA, A. B., HUANG, K., HUBY, R. and NEWNS, H. C., *Phil. Mag.* **43**, 485 (1952).
BJERGE, T. and WESTCOTT, C. H., *Proc. Roy. Soc. (London)* **A150**, 709 (1935).
BJORKLUND, F. and FERNBACH, S., *Phys. Rev.* **109**, 1295 (1958).
BLAIR, J. S., *Phys. Rev.* **115**, 928 (1959).
BLAIR, J. S., in *Proceedings of the International Conference on Nuclear Structure, Kingston*, p. 824, edited by D. A. Bromley and E. W. Vogt, University of Toronto Press, Toronto, 1960.
BLATT, J. M. and WEISSKOPF, V. F., *Theoretical Nuclear Physics*, John Wiley, New York, 1952.
BLOCH, C., *Nucl. Phys.* **4**, 503 (1957).
BOHR, N., *Nature* **137**, 344 (1936).
BOHR, N. and KALCKAR, F., *Kgl. Danske Videnskab. Selskab, Mat-fys. Medd.* **14**, No. 10 (1937).
BOHR, A. and MOTTELSON, B. R., *Kgl. Danske Videnskab. Selskab, Mat-fys. Medd.* **27**, No. 16 (1953, 2nd edition 1957).
BREIT, G. and WIGNER, E. P., *Phys. Rev.* **49**, 519 (1936).
BRINK, D. M., *Nuclear Forces*, Pergamon Press, London, 1965.
BRINK, D. M. and STEPHEN, R. O., *Phys. Letters* **5**, 77 (1963).
BRUECKNER, K. A., LOCKETT, A. M. and ROTENBERG, M., *Phys. Rev.* **121**, 255 (1961).
BUCK, B., *Phys. Rev.* **130**, 712 (1963).

BURROWS, H. B., GIBSON, W. M. and ROTBLAT, J., *Phys. Rev.* **80**, 1095 (1950).
BUTLER, S. T., *Phys. Rev.* **80**, 1095 (1950); *Proc. Roy. Soc. (London)* **A208**, 559 (1951).
CARLSON, A. D. and BARSCHALL, H. H., *Phys. Rev.* **158**, 1142 (1967).
CHADWICK, J., *Proc. Roy. Soc. (London)* **A136**, 692 (1932).
CHADWICK, J. and BIELER, E. S., *Phil. Mag.* **42**, 923 (1921).
CHADWICK, J., CONSTABLE, J. E. R. and POLLARD, E. C., *Proc. Roy. Soc. (London)* **A130**, 463 (1931).
CHAMBERLAIN, O. and SEGRÈ, E., *Phys. Rev.* **87**, 81 (1952).
CHASE, D. M., WILETS, L. and EDMONDS, A. R., *Phys. Rev.* **110**, 1080 (1958).
COHEN, B. L., *Phys. Rev.* **92**, 1245 (1953).
COHEN, B. L., *Phys. Rev.* **108**, 768 (1957).
COHEN, B. L., *Am. J. Phys.* **33**, 1011 (1965).
COOK, L. J., MCMILLAN, E. M., PETERSON, J. M. and SEWELL, D. C., *Phys. Rev.* **75**, 7 (1949).
DODD, L. R. and MCCARTHY, I. E., *Phys. Rev.* **134**, A1136 (1964).
DROZDOV, S. I., *Zh. Eksperim. i Teor. Fiz.* **28**, 734, 736 (1955) [English translation: *Soviet Phys. JETP* **1**, 591, 588 (1955)].
EISBERG, R. M. and IGO, G. J., *Phys. Rev.* **93**, 1039 (1954).
EISBERG, R. M., MCCARTHY, I. E. and SPURRIER, R. A., *Nucl. Phys.* **10**, 571 (1959).
EL BEDEWI, F. A., MIDDLETON, R. and TAI, C. T., *Proc. Phys. Soc. (London)* **A64**, 756, 1055 (1951).
ELLIOTT, J. P. and LANE, A. M. The nuclear shell model, *Encyclopedia of Physics*, Vol XXXIX, Berlin, Springer-Verlag, 1957.
ELSASSER, W. M., *J. Phys. et Radium* **5**, 389, 635 (1934).
FERNBACH, S., SERBER, R. and TAYLOR, T. B., *Phys. Rev.* **75**, 1352 (1949).
FESHBACH, H., PEASLEE, D. C. and WEISSKOPF, V. F., *Phys. Rev.* **71**, 145 (1947).
FESHBACH, H., PORTER, C. E. and WEISSKOPF, V. F., *Phys. Rev.* **96**, 448 (1954).
FEYNMAN, R. P., *Phys. Rev.* **76**, 749, 769 (1949).
FRISCH, O. R. and PLACZEK, G., *Nature* **137**, 357 (1936).
GILLET, V. and VINH MAU, N., *Nucl. Phys.* **54**, 321 (1964).
GLASSGOLD, A. E. and KELLOGG, P. J., *Phys. Rev.* **109**, 1291 (1958).
GOMES, L. C., WALECKA, J. D. and WEISSKOPF, V. F., *Ann. Phys. (N.Y.)* **3**, 241 (1958).
GOODING, T. J. and PUGH, H. G., *Nucl. Phys.* **18**, 46 (1960).
GOSWAMI, A. and PAL, M. K., *Nucl. Phys.* **44**, 294 (1963).
GRIFFITHS, R. J. and EISBERG, R. M., *Nucl. Phys.* **12**, 225 (1959).
GUGELOT, P. C., *Phys. Rev.* **93**, 425 (1954).
GURNEY, R. W., *Nature* **123**, 565 (1929).
HAUSER, W. and FESHBACH, H., *Phys. Rev.* **87**, 366 (1952).
HAXEL, O., JENSEN, J. H. D. and SUESS, H. E., *Phys. Rev.* **75**, 1766 (1949); *Z. Phys.* **128**, 295 (1950).
HAYBRON, R. M., *Nucl. Phys.* **79**, 33 (1966).
HAYBRON, R. M. and MCMANUS, H., *Phys. Rev.* **140**, B638 (1965).
HEISENBERG, W., *Z. Phys.* **96**, 473 (1935).
HELMHOLZ, A. C., MCMILLAN, E. M. and SEWELL, D. C., *Phys. Rev.* **72**, 1003 (1947).

HOFSTADTER, R., *Rev. Mod. Phys.* **28**, 214 (1956).
HOLT, J. R. and YOUNG, C. T., *Proc. Phys. Soc.* (*London*) **A63**, 833 (1950).
HUMBLET, J. and ROSENFELD, L., *Nucl. Phys.* **26**, 529 (1961).
INOPIN, E. V., *Zh. Eksperim. i Teor. Fiz.* **31**, 901 (1956) [English translation: *Soviet Phys. JETP* **4**, 764 (1957)].
JACOB, G. and MARIS, TH. A. J., *Rev. Mod. Phys.* **38**, 121 (1966).
JAIN, B. K. and JACKSON, D. F., *Nucl. Phys.* **A99**, 113 (1967).
JOHNSON, R. R. and HINTZ, N. M., *Phys. Rev.* **153**, 1169 (1967).
KAPUR, P. L. and PEIERLS, R. E., *Proc. Roy. Soc.* (*London*) **A166**, 277 (1938).
KENDALL, M. G., *The Advanced Theory of Statistics*, Charles Griffin and Company, Ltd., London, 1946.
KÖHLER, H. S., *Nucl. Phys.* **88**, 529 (1966).
KROMMINGA, A. J. and MCCARTHY, I. E., *Nucl. Phys.* **31**, 678 (1962).
LAWRENCE, E. O. and LIVINGSTONE, M. S., *Phys. Rev.* **40**, 19; **42**, 150 (1932).
LAWRENCE, E. O., LIVINGSTONE, M. S. and LEWIS, G. N., *Phys. Rev.* **44**, 56 (1933).
LEE, L. L., JR., SCHIFFER, J. P., ZEIDMAN, B., SATCHLER, G. R., DRISKO, R. M. and BASSEL, R. H., *Phys. Rev.* **136**, B971 (1964).
LEVINSON, C. A. and BANERJEE, M. K., *Ann. Phys.* (*N.Y.*) **2**, 471, 499 (1957); **3**, 67 (1958).
LIM, K. L. and MCCARTHY, I. E., *Phys. Rev.* **133**, B1006 (1964).
LIM, K. L. and MCCARTHY, I. E., *Nucl. Phys.* **88**, 433 (1966).
MCCARTHY, I. E., *Rev. Mod. Phys.* **37**, 388 (1965).
MCCARTHY, I. E. *Nucl. Phys.* **10**, 583; **11**, 574 (1959).
MCCARTHY, I. E. and PURSEY, D. L., *Phys. Rev.* **122**, 578 (1961).
MCDANIELS, D. K., BLAIR, J. S., CHEN, S. Y. and FARWELL, G. W., *Nucl. Phys.* **17**, 614 (1960).
MACFARLANE, M. H. and FRENCH, J. B., *Rev. Mod. Phys.* **32**, 567 (1960).
MAYER, M. G., *Phys. Rev.* **74**, 235 (1948); **75**, 1969 (1949); **78**, 16, 22 (1950).
MAYER, M. G. and JENSEN, J. H. D., *Elementary Theory of Nuclear Shell Structure*, John Wiley, New York, 1955.
MEHTA, M. L., *Nucl. Phys.* **18**, 395 (1960).
MILLER, D. W., ADAIR, R. K., BOCKELMAN, C. K. and DARDEN, S. E., *Phys. Rev.* **88**, 83 (1952).
MOON, P. B. and TILLMAN, J. R., *Proc. Roy. Soc.* (*London*) **A153**, 476 (1936).
MOTT, N. F., *Proc. Roy. Soc.* (*London*) **A133**, 228 (1931).
MOSZKOWSKI, S. A., Models of nuclear structure, *Encyclopedia of Physics*, Vol. XXXIX, Berlin, Springer Verlag, 1957.
MOSZKOWSKI, S. A. and SCOTT, B. L., *Ann. Phys.* (*N.Y.*) **11**, 65 (1960); **14**, 107 (1961); *Nucl. Phys.* **29**, 665 (1962).
OPPENHEIMER, J. R. and PHILLIPS, M., *Phys. Rev.* **47**, 845; **48**, 500 (1935).
PAL, M. K. and STAMP, A. P., *Phys. Rev.* **158**, 924 (1967).
PALEVSKY, H., FRIEDES, J. L., SUTTER, R. J., BENNETT, G. W., IGO, G. J., SIMPSON, W. D., PHILLIPS, G. C., CORLEY, D. M., WALL, N. S., STEARNS, R. L. and GOTTSCHALK, B., *Phys. Rev. Letters* **18**, 1200 (1967).
PAULING, L. and WILSON, E. B., *Introduction to Quantum Mechanics*, McGraw-Hill, New York, 1935.

Perey, F. G., *Phys. Rev.* **131**, 745 (1963).
Perey, F. G. and Buck, B., *Nucl. Phys.* **32**, 353 (1962).
Porter, C. E. and Thomas, R. G., *Phys. Rev.* **104**, 483 (1956).
Pugh, H. G., Hendrie, D. L., Chabre, M. and Boschitz, E., *Phys. Rev. Letters* **14**, 434 (1965).
Pugh, H. G., Hendrie, D. L., Chabre, M., Boschitz, E. and McCarthy, I. E., *Phys. Rev.* **155**, 1054 (1967).
Riezler, W., *Proc. Roy. Soc. (London)* **134**, 154 (1932).
Riou, M., *Rev. Mod. Phys.* **37**, 375 (1965).
Roberts, R. B. and Abelson, P. H., *Phys. Rev.* **72**, 76 (1947).
Rost, E. and Austern, N., *Phys. Rev.* **120**, 1375 (1960).
Rutherford, E., *Proc. Roy. Soc. (London)* **A123**, 323 (1929).
Schiff, L. I., *Quantum Mechanics*, McGraw-Hill, New York, 1955.
Serber, R., *Phys. Rev.* **72**, 1114 (1947).
Serber, R., *Phys. Rev.* **72**, 1008 (1947a).
Stehn, J. R., Goldberg, M. D., Wiener-Chasman, R., Mughabghab, S. F., Magurno, B. A. and May, V. M., Brookhaven National Laboratory Report No. BNL 325, 2nd edition, Supplement No. 2, 1965 (unpublished).
Szilard, L., *Nature* **136**, 950 (1935).
Tamura, T., *Rev. Mod. Phys.* **37**, 679 (1965).
Tamura, T. and Terasawa, T., *Prog. Theor. Phys.* **26**, 285 (1962).
Taylor, H. M., *Proc. Roy. Soc. (London)* **136**, 605 (1932).
Thomas, L. H., *Phys. Rev.* **54**, 580 (1938).
Tyrén, H., Hillman, P. and Maris, Th. A. J., *Nucl. Phys.* **7**, 10 (1958).
Tyrén, H., Hillman, P. and Maris, Th. A. J., *Phys. Rev. Letters* **5**, 107 (1960).
Veksler, V. I., *J. Phys. U.S.S.R.* **9**, 153 (1945).
Vogt, E. W., McPherson, D., Kuehner, J. and Almquist, E., *Phys. Rev.* **136**, B99 (1964).
Walt, M. and Barschall, H. H., *Phys. Rev.* **93**, 1062 (1954).
Watson, K. M., *Rev. Mod. Phys.* **30**, 565 (1958).
Weisskopf, V. F. and Ewing, D. H., *Phys. Rev.* **57**, 472, 935 (1940).
Weizsäcker, C. F. von, *Z. Phys.* **96**, 431 (1935).
Wheeler, J. A., quoted by Bethe (1937).
Wigner, E. P., *Am. J. Phys.* **23**, 371 (1955).
Wigner, E. P. Gatlinburg Conference on Neutron Physics by Time of Flight, p. 59. Oak Ridge National Laboratory Report No. ORNL-2309, 1957 (unpublished).
Wigner, E. P. and Eisenbud, L., *Phys. Rev.* **72**, 29 (1947).
Wolfenstein, L., *Phys. Rev.* **82**, 690 (1951).
Woods, R. D. and Saxon, D. S., *Phys. Rev.* **95**, 577 (1954).
Wyatt, P. J., Wills, J. G. and Green, A. E. S., *Phys. Rev.* **119**, 1031 (1960).

PART 2

1

Discussion on the Structure of Atomic Nuclei†

SIR ERNEST RUTHERFORD

SIR ERNEST RUTHERFORD: It was on March 19, 1914, that the Royal Society held its last discussion on the constitution of the atom—just fifteen years ago. I had the honour to open the discussion on that occasion, and the other speakers were Mr. Moseley, Profs. Soddy, Nicholson, Hicks, Stanley Allen, S. P. Thompson. In my opening remarks I put forward the theory of the nuclear atom and the evidence in support of it, while Mr. Moseley gave an account of his X-ray investigations, which defined the atomic numbers of the elements, and showed how many gaps were present between hydrogen number 1 and uranium number 92. Prof. Soddy drew attention to the existence of isotopes in the radioactive series, and also to a remarkable observation by Sir Joseph Thomson and Dr. Aston, who had obtained two parabolas in the positive ray spectrograph of neon, and he suggested that possibly the ordinary elements might also consist of mixture of isotopes. I think you will find that the remarks and suggestions made in this discussion fifteen years ago have a certain pertinence to-day. In particular Hicks and Stanley Allen drew attention to the importance of taking into account the magnetic fields in the nucleus, although at that time we had very little evidence on that point, and even to-day our information is very scanty.

What has been accomplished in the intervening period? On looking back we see that three new methods of attack on this problem have been developed. The first, and in some respects the

† *Proc. Roy. Soc.* **A123**, 373 (1929).

most important, has been the proof of the isotopic constitution of the ordinary elements, and the accurate determination of the masses or weights of the individual isotopes, mainly due to the work of Dr. Aston. This has led in a sense to an extension of the original ideas of Moseley. The experiments of the latter fixed the number of possible nuclear charges, while Aston has shown that there are a large number of species of atoms each defined by its nuclear charge, although their masses and their nuclear constitution may be different. The essential point brought out in the earlier work of Dr. Aston was that the masses of the elements are approximately expressed by whole numbers, where oxygen is taken as 16— with the exception of hydrogen itself. But the real interest, as we now see it, is not the whole number rule itself, but rather the departures from it. That is a point I will discuss at some length later. It suffices to say at this point that the evidence from isotopes and also from the artificial disintegration of light elements shows fairly clearly that the mass units entering into the composition of the nucleus have a mass in the nucleus of about 1. This unit has been named the proton, and we believe that the proton is identical with the hydrogen nucleus when in the free state. I pointed out in the 1914 discussion that almost certainly the hydrogen nucleus corresponded to the positive electron—the counterpart of the ordinary negative electron.

The next discovery was the proof of the artificial disintegration of the elements by bombardment with alpha particles, in which I have been personally interested, as has also Dr. Chadwick, who will discuss the results obtained and their bearing on the structure of nuclei. This is the first time, I think, that definite evidence has been obtained that we can alter the actual structure of the nucleus itself by the application of external agencies. We know that in all cases where this is effected a proton is liberated at high speed. It is noteworthy that while the radioactive substances spontaneously break up always with the emission of helium nuclei or electrons, in the artificial disintegration of the lighter atoms, no helium nuclei, as far as we know, are liberated, but a proton is set free. The evidence, as a whole, indicates that while the ultimate constituents of the nucleus are protons and electrons, secondary units in the form of helium nuclei are also present in the heavier elements. I shall refer later to this important question.

The third line of attack, which has proved of great interest, has been the study of the wave-lengths of the penetrating gamma-rays which arise during the disintegration of the radioactive nucleus. As the gamma-rays have their origin in the nucleus, the frequency of these rays, first determined by Dr. Ellis, give us important evidence as to the modes of vibration, to use a general term, of the particles constituting the nucleus. I hope that Dr. Ellis later will refer among other matters to certain problems of the relation of the constituents of the nucleus with the outer electrons on which he has obtained at any rate some tentative evidence.

Dr. Chadwick and I have been especially interested in recent years in the question of the dimensions of the nucleus and the laws of force which hold in its neighbourhood. This is highly important since it is evident that we can make no calculations of quantities until we know the nature and laws of the forces near a nucleus. Information of this can be obtained by a study of the scattering of α-particles, and we have performed a large number of experiments on different elements. The methods adopted are in principle extremely simple. An intense beam of α-particles of definite speed falls on a thin sheet of matter and the number of α-particles scattered through an angle of about 135° is counted by the scintillation method. The number of α-particles observed in this way is, in general, about 1 in 10^5 of the α-particles incident on the scattering foil. The speed of the incident α-particles is varied by placing thin sheets of mica over the source. In this way, a determination is made of the number of particles scattered by α-particles of different speeds. If the law of electrostatic action between the nucleus and the α-particle is the ordinary law of inverse square, the number scattered ought to vary as $1/E^2$, where E is the energy of the α-particle. When we examine all elements, from copper of atomic number 29 to uranium of number 92, it is found that the scattering is normal; that is to say, that the scattering agrees within experimental error with what we should expect if the law of inverse square holds in the region which is penetrated by the α-particle. Since it is to be anticipated that the law of force would change if the α-particle penetrates a nucleus, we may conclude that the radius of the copper nucleus must be less than the closest distance of approach, which in this case is about 10^{-12} cm. The corresponding distance for the uranium nucleus is about 3×10^{-12} cm, using the swiftest α-particles available. Since no change in the law of force is observed over this wide range of atomic number, we are unable to fix the dimensions of the nucleus with any certainty. All we can say is that it must be smaller than the closest distance of approach in a collision between the α-particle and the nucleus in question. If the α-particle entered into the nucleus, it is to be expected that the forces would alter, resulting in a change of the law of scattering.

Quite a different result is observed when we examine the scattering of the lighter elements. Long ago it was found that in bombarding hydrogen with α-particles the scattering is completely abnormal, and a similar result has been recently observed for helium. A detailed examination, first by Bieler and later by Chadwick and myself, has been made of the scattering of α-particles by magnesium and aluminium of atomic numbers 12 and 13 respectively. I will not go into the details of the experiments, but will merely illustrate the type of scattering curve which is observed.

Let us first suppose that the scattering is normal, that is, that the number of α-particles scattered through an angle of 135° varies as $1/E^2$ where E is the energy of the α-particle. The ratio of the observed to the theoretical scattering for different velocities of the α-particle is given by the dotted line shown in the figure. Such a straight line is found for gold and all the intermediate elements examined down to copper; but when aluminium is the scattering material, it is found that the inverse square law is nearly true for slow particles. For increasing velocity of the α-particle, the curve falls below the normal line, passes through a minimum and then rises again. There is some evidence that if still swifter α-particles were available, the curve would rise steeply well above the normal line. The curve shown in the figure is for the scattering of α-particles by aluminium through an average angle of 135°. A similar curve is found for magnesium. It is probable that still lighter elements would show a similar variation in the scattering curve. In some respects, the scattering of α-particles by hydrogen and helium is closely analogous.

The most natural explanation, and one which I think fits in well with the results, is to suppose that, in addition to the ordinary repulsive forces of the electric type between the nuclei, at close distances attractive forces come into play, so that the resultant force is a combination of the forces of repulsion and attraction.

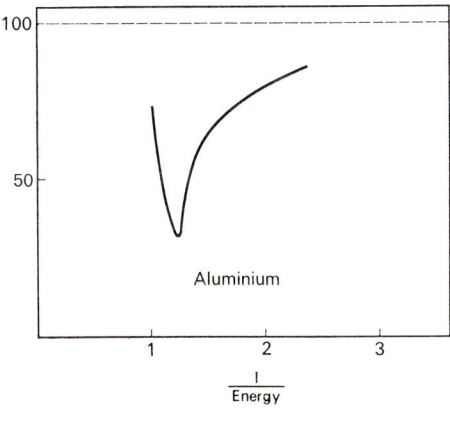

Fig. 1.

The scattering to be expected under these conditions was first worked out by Bieler, taking the attractive force to vary as the inverse fourth power of the distance. Later Debye and Hardmeier worked out the case for the inverse fifth power—a law of attraction that has a certain physical significance. When an α-particle approaches near to a nucleus, the intense forces must distort or polarize the charged constituents of the nucleus. This should give rise to an attractive force on the colliding α-particle, varying approximately as the fifth power of the distance from the centre of the nucleus. This point of view, it seems to me, has much to recommend it, for there must be a distortion or polarization of the nucleus under the action of the intense forces, and this resulting attraction may become very great when the α-particle approaches near to the nucleus.

It has been shown by Hardmeier that calculations on this basis are in fair accord with the experimental observations for aluminium. It should be pointed out that the assumptions made in these calculations are somewhat artificial, for the α-particle is supposed to be a point charge and the nucleus a sphere.

Observations of the scattering of α-particles by hydrogen and helium indicate that both the hydrogen and helium nucleus appear to be surrounded by a field of force of unknown origin where the laws of force are quite abnormal. This region of abnormal forces is of about the same dimensions as the aluminium nucleus and the interpretation of the evidence suggests that in collisions this region is not spherical but rather like a plate or flat ellipsoid. It is of interest to note that the "size" of the hydrogen nucleus or proton examined in this way appears to be even larger than that of the helium nucleus.

Whatever interpretation we may place on the experimental results, it is clear that these nuclei cannot be regarded as point forces but have a certain volume or structure. It has been suggested, and it may be true, that this peculiar distribution of the forces round the α-particle and the proton may be due to magnetic forces which, on modern views, may arise from the intrinsic magnetic moment ascribed to the proton, and possibly also the electron, quite apart from the actual motions of the constituents of the nucleus.

Now I come to an important point of the development of this argument. The observations on the scattering of α-particles by uranium metal show that within experimental error—which is unfortunately rather large in this case—the scattering is quite normal for the closest distance of approach—about $3 \cdot 5 \times 10^{-12}$ cm—indicating that the radius of the nucleus is still smaller than this quantity. At the same time, if we examine the speed with which the slowest α-particle is spontaneously emitted from the uranium

DISCUSSION ON THE STRUCTURE OF ATOMIC NUCLEI 131

nucleus, we are driven to the conclusion that the nucleus extends to a distance about $6 \cdot 5 \times 10^{-12}$ cm—about twice the distance found from the scattering results. We are thus faced with the difficulty that two apparently reliable methods of estimating the size of nuclei give very different values. If we try to construct a nucleus on classical ideas and to make the model self-consistent, we are driven to the conclusion that whatever system of forces is adopted, the field of force round the uranium nucleus must consist of an attractive force at small distances and a repulsive force at large distances. The variation of the potential due to these forces is shown in the diagram. The maximum of the curve represents the

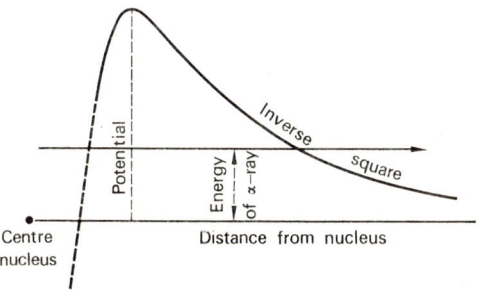

Fig. 2.

distance where the attractive and repulsive forces balance each other. Inside this distance the attractive forces preponderate and the potential may fall to a negative value. At large distances, the potential due to the electrostatic forces varies inversely as the distance. I think that all will agree that the potentials in the neighbourhood of the nucleus must vary somewhat in this fashion. On classical dynamics the potential may have a value of 4 or 5 million volts at a distance 4 to 5×10^{-12} cm about. Whatever types of forces are adopted, a similar result follows. We are driven then, on these views, to the conclusion that a swift α-particle can penetrate deeply into the uranium nucleus while the scattering evidence contradicts such an assumption.

There have been attempts during the past year to avoid this impasse, and to get further information by applying the ideas of the wave mechanics. This problem has been attacked by Mr. Gamow, whom we are very glad to welcome here to-day, and also by Gurney and Condon. The variation of potential near the nucleus is supposed to be very similar in shape to that shown in Fig. 2, but from the calculations of Mr. Gamow, the peak of the curve is much closer to the nucleus, at about 0.7×10^{-12} cm instead of 4 or 5×10^{-12} cm. The maximum potential is correspondingly higher, about 30 million volts, and falls off exceedingly rapidly close to the nucleus. The nucleus, supposed spherical in shape, is thus surrounded by a very high potential barrier. No α-particle born of uranium can jump this barrier for, if it did so, it would escape with an energy much greater than is observed for any α-particle. But on the wave mechanics, particles can accomplish feats that are quite impossible on classical mechanics. Instead of the α-particle being required to jump the barrier in order to escape, the particle, or rather the wave system with which it is identified, leaks out through the barrier and finally emerges with a kinetic energy equal to the total energy it possessed inside the barrier. I will not deal further with this interesting new point of view, which will be discussed later by Mr. Fowler and Mr. Gamow. It will be seen that this theory makes the radius of the uranium nucleus very small, about 7×10^{-13} cm, and in this small nuclear volume 238 protons and 146 electrons have to be made room for. It sounds incredible but may not be impossible.

The next point I will consider is the bearing of some of the results of Dr. Aston on the structure of nuclei. You all know the essential result at which he had arrived—that the unit of mass in the structure of nuclei is the proton of mass nearly 1, while the proton (H nucleus) in the free state has a mass 1.0073 in terms of $O = 16$. This difference in mass between the free and nuclear proton is ascribed to this packing effect, i.e., to the interaction of the electromagnetic fields of the protons and electrons in the highly condensed nucleus. On modern views, we know that there is a close relation between mass and energy and a decrease in mass means a loss of energy. The free proton has a mass 1·0073 while the proton in the nucleus has a mass very nearly 1. This apparently small loss of mass means that a large amount of energy has been emitted in the entrance of the free proton into the structure of the nucleus, an amount corresponding to about 7 million electron-volts.

Consider now the curve in Fig. 3 prepared by Dr. Aston which shows on a large scale the departure of the masses of the isotopes from the whole number rule. The curve beginning at hydrogen crosses the whole number line and sinks to a minimum for an

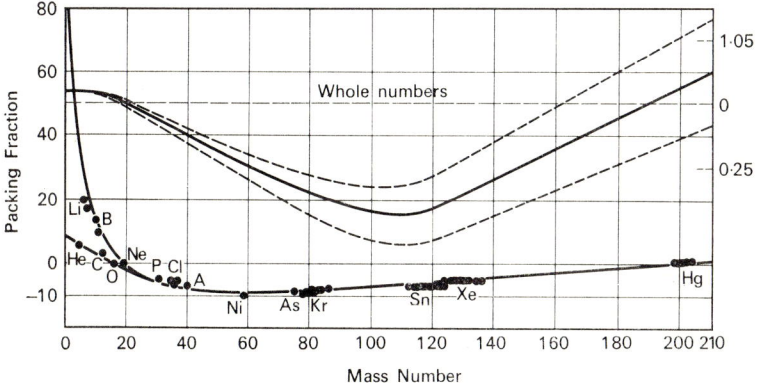

FIG. 3. Mass numbers and packing fractions.

atomic weight about 120. It then rises and crosses the line again for an atomic weight about 200. Suppose now that a nucleus is built up originally from free hydrogen nuclei and electrons—I do not say I agree with such an assumption. In forming an atom of mass 120, containing 120 protons, energy has been lost corresponding to 840 million volts, and this energy has been probably radiated away into space. On such a view it is clear that the atom, instead of being as ordinarily supposed, a store of energy, is really a sink of energy; and if we took the nucleus of mercury, for example, to pieces, proton by proton, we should have to do an enormous amount of work corresponding to at least 1400 million volts.

This is certainly true if each atom is supposed to be built up of free hydrogen nuclei and electrons. You may naturally ask how such a view can be reconciled with the fact that in the transformation of uranium into lead a large amount of energy is spontaneously liberated—more than 40 million volts. If, however, we look at it from another point of view, the prospect is more encouraging. Suppose that the essential unit involved in the structure of the heavier elements is not the proton but the helium nucleus. The latter which we suppose to be made up of 4 protons and 2 electrons has already in its formation emitted a large amount of energy and its nuclear mass in the free state is 4·0018. It will be noted that the curve showing the departure from the whole number rule is a minimum for a mass of about 120 and crosses the line again for a mass of about 200. If, for simplicity, we suppose the nuclei are built up from mass 120 by successive accretions of α-particles, the mass of each added particle can be easily shown to be about 4·005 over this interval of atomic weight. It is seen that this mass is greater than that of the α-particle in the free state. No doubt also the increase of mass is not regular from atom to atom and is probably greater for a mass in the neighbourhood of 200 than for a mass about 120. If the curve given by Aston is extrapolated to include uranium, it can be shown that the addition of mass is sufficient to account at any rate approximately for the large amount of energy emitted by the radioactive bodies. It appears therefore very likely that the heavier elements are mainly composed of α-particles and that many of the α-particles have a total energy in the nucleus greater than in the free state.

We are now in a position to form a picture of the gradual building up of atomic nuclei. Probably in the lighter elements the nucleus is composed of a combination of α-particles, protons, and electrons, and that the parts of the nucleus attract one another strongly, partly it may be owing to the distortional forces and partly also to the magnetic forces. We can only speculate as to the nature of these forces. At first then a highly concentrated and firmly bound nucleus is formed accompanied by the emission of energy and for an atomic weight of about 120 a minimum mass is reached representing the closest binding. After this stage, as the atom grows in mass, the additional particles are less and less tightly bound. We may thus suppose that the nucleus consists of a tightly packed structure near its centre gradually becoming less dense towards the outside. This system is surrounded by a potential barrier which normally prevents the α-particles from escaping. This static view of the atom may not commend itself to my theoretical friends who may wish the α-particles to have complete freedom of motion within the nucleus. Such a point of view is quite legitimate and can be reconciled with the essential ideas I have put forward. In other words, if it were possible to take an instantaneous picture of the nucleus—of duration about 10^{-23} sec—a dense collection of closely bound α-particles would be observed at the centre, the density fading away to the outside. No doubt all the α-particles are in motion and their waves are reflected to and fro between the potential barriers, and occasionally leak out of the system. The view I have outlined is I think reasonable, and I hope our theoretical friends will be able to define the picture more accurately; we have not only to account for a nucleus made of α-particles but have also to make room for the additional electrons, and it is not easy to confine an electron in the same cage with an α-particle. I have, however, such a strong belief in the ingenuity of our theoretical friends that I am confident that they will surmount this difficulty in some way and I hope suggestions will be made on this point to-day.

One other point. The view I have put forward explains I think why no heavier atoms than uranium can exist permanently. With increase of mass, the nuclei gain more and more energy and would become highly radioactive and disappear. Probably the higher the energy the more rapidly they would vanish, and it is probably not an accident that uranium and thorium are the sole survivors of the heavier nuclei. This is not the place to discuss the very speculative question of how the nuclei of the elements have been formed. Before attacking this question, we want much more information on the details of the structure of the nucleus itself.

2

The Theory of the Effect of Resonance Levels on Artificial Disintegration[†]

N. F. MOTT

Section 1. Discussion of Previous Investigations

According to the theory first proposed by Gamow, and by Gurney and Condon, the emission of α-particles by radio active nuclei is to be explained by the assumption that there exists in the nucleus a "virtual" level of positive energy, which is occupied by an α-particle. According to the wave mechanics, such levels are necessarily unstable, and there exists a definite probability that the α-particle will escape in a given interval of time. The theory has been discussed by a great many authors; we shall need to refer in particular to the calculations of Fowler and Wilson,[‡] who have given the exact wave function that describes the emission from a simplified one-dimensional nucleus.

We shall need the following result, which may be found in Fowler and Wilson's paper or elsewhere. Let τ be the natural period of the α-particle in the nucleus, i.e., the time taken for a particle moving with velocity of about 10^9 cm/sec to travel a distance of about 3×10^{-13} cm, so that

$$\tau \sim 10^{-22} \text{ sec.}$$

Let T be the time of decay of an α-particle in a given level. Let us define a quantity ζ^2 by the equation

$$\zeta^2 = T/\tau.$$

[†] *Proc. Roy. Soc.* **A133**, 228 (1931). Communicated by P. A. M. Dirac.
[‡] *Proc. Roy. Soc.* **A124**, 493 (1929).

EFFECT OF RESONANCE LEVELS   137

Then the α-particles emitted by the nucleus will have their energies distributed in a range ΔE given by

$$\frac{\Delta E}{E} \sim \frac{1}{\zeta^2}. \tag{1}$$

The quantity ζ^2 is thus a measure of the breadth of the virtual level.

Now it was first pointed out by Gurney[†] that there might exist in the lighter nuclei similar virtual energy levels that are *not* normally occupied by an α-particle, and that if a nucleus were bombarded by α-particles having just the energy of one of these levels, the probability that an α-particle would penetrate into the nucleus would be much greater than would be the case for α-particles of other energies. It would follow that the number of H-particles produced should show strong maxima for certain velocities of the incident α-particle. It is, of course, supposed that these virtual levels are much broader than those that occur in the radioactive nuclei, the quantity ζ^2 being, perhaps, of the order of magnitude 50; for a radioactive element ζ^2 is 10^{20} or 10^{30}. Recently the existence of such "resonance" levels has been demonstrated by Pose, who finds levels in aluminium at $3 \cdot 9 \times 10^6$ and $4 \cdot 8 \times 10^6$ volts. The breadth of these levels is not known; we shall make an estimate of it in section 4.

In this paper we shall be concerned with the effect that the resonance levels may be expected to have on the artificial disintegration and on the anomalous scattering. We shall express our results in terms of cross-sections, calculating the effective area that an α-particle must hit to eject a proton, being itself captured, and the area that it must hit in order that it shall be scattered into a given solid angle $d\omega$.

We shall denote the cross-section for capture by $a(E)$, and the cross-section for scattering into a solid angle $d\omega$ by $I_\theta d\omega$. We shall investigate the circumstances under which a resonance level should occur, and estimate its breadth ΔE. We shall find an expression (equation (15)) for the maximum value of $a(E)$ within the resonance level. We shall also investigate the deviations of I_θ from the

[†] *Nature*, **123**, 565 (1929).

inverse square form, $(ZZ'\epsilon^2/2mv^2 \sin^2 \tfrac{1}{2}\theta)^2$, that are to be expected for energies of the incident α-particle lying in the resonance range. None of the results of this paper depend on the form assumed for the interaction between the α-particle and the proton.

The effect of resonance levels on artificial disintegration has been considered by Fowler and Wilson,[†] Atkinson,[‡] and Guido Beck.[§] The first two authors discuss a one-dimensional nucleus, and they consider only the probability that the α-particle will be captured, i.e. that it will remain in the nucleus after collision for a time comparable with the time of decay. They do not investigate the probability of a transference of energy from the α-particle to a proton. They find that if the α-particle has energy equal to a resonance level of the nucleus, the probability of capture is of the order of magnitude unity; otherwise it is negligible. They do not generalize their results to apply to three degrees of freedom.

Beck has investigated the probability that the α-particle should lose energy to a proton, and thus be captured into a stable level. He assumes a Coulomb interaction between the α-particle and the proton, and obtains the probability for capture by treating this interaction as small. This method, however, gives misleading results unless the interaction energy is very small, as we shall see in this paper.

Section 2

We shall now consider what will happen when a beam of α-particles falls on a nucleus. Let the potential energy $V(r)$ of the α-particle in the field of the nucleus be a function of the usual type, giving Coulomb repulsion at large distances and attraction at small distances. We assume that this field has a virtual level E_0 of breadth[††] E_0/ζ^2. We assume that for energies just above and below this range the chance of penetration is negligible. We

---

[†] *Loc. cit.*
[‡] *Z. Physik*, **64**, 507 (1930).
[§] *Z. Physik*, **64**, 22 (1930).
[††] The order of magnitude of ζ^2 may be about 100.

EFFECT OF RESONANCE LEVELS 139

assume further that E_0 is the energy of an S level, i.e., that the wave function associated with it is spherically symmetrical.

Let us consider first what will happen if the α-particle has not enough energy to eject a proton, or to knock it into an excited state. Then, since all particles eventually escape, all that could be observed would be anomalous scattering. This we shall now investigate.

The wave equation for an α-particle in the field $V(r)$ is

$$\nabla^2 \psi + \frac{8\pi^2 m}{h^2}(E - V)\psi = 0. \tag{2}$$

The general solution of (2) is

$$\psi = \sum_{n=0}^{\infty} A_n P_n(\cos\theta) L_n(r), \tag{3}$$

where the A_n are arbitrary constants, and L_n is the solution, bounded at the origin, of

$$\frac{d^2L}{dr^2} + \frac{2}{r}\frac{dL}{dr} + \left\{\frac{8\pi^2 m}{h^2}(E-V) - \frac{n(n+1)}{r^2}\right\}L = 0. \tag{4}$$

The constants A_n must be chosen so that (3) represents an incident wave of unit amplitude, together with a scattered wave. This can be done by the method of Faxén and Holtzmark[†] for any given field $V(r)$.

For the case of the Coulomb field, when $V(r)$ is equal to $ZZ'\epsilon^2/r$, the constants A_n have been found by Gordon.[‡] We shall need certain of his results, and give them here, writing L_n^c, A_n^c, etc., for the forms that these functions take for this particular $V(r)$. If $L_n^c(r)$ be suitably normalized, it has the asymptotic form

$$L_n^c(r) \sim r^{-1}\cos(kr - (n+1)\pi/2 - \alpha\log 2kr + \sigma_n), \tag{5}$$

[†] Faxén and Holtzmark, *Z. Physik*, **45**, 307 (1927).
[‡] Gordon, *Z. Physik*, **48**, 180 (1928).

where

$$k = 2\pi mv/h = 2\pi\sqrt{2mE}/h,$$

$$a = 2\pi ZZ'\epsilon^2/hv = \frac{2Z}{137}\frac{c}{v},$$

$$\sigma_n = \arg \Gamma(ia + n + 1).$$

Gordon finds for $A_n{}^c$

$$A_n{}^c = \frac{1}{k} i^n (2n + 1) e^{i\sigma_n} \tag{6}$$

and finds for (3) a wave function ψ^c with the asymptotic form

$$\psi^c \sim I + f^c(\theta) S,$$

where I the incident wave is given by

$$I = \exp(ikz + ia \log k(r - z))$$

and for the scattered we have

$$S = \exp(ikr - ia \log 2kr)$$

$$f^c(\theta) = \frac{2Z\epsilon^2}{2mv^2} \operatorname{cosec}^2 \tfrac{1}{2}\theta \exp(-ia \log \sin^2 \tfrac{1}{2}\theta + 2i\sigma_0 + i\pi).$$

Now if E does not lie in the neighbourhood of E_0, then the solution of the equation (4) for the actual nuclear field $V(r)$ will be practically the same as for a Coulomb field. If, on the other hand, E lies in the resonance range, then this will still be the case, except when $n = 0$. This follows from our assumption that the virtual level is an S level. If we denote the solution of (4) for the case $n = 0$ by $L_0{}^N(r)$, then the asymptotic form of $L_0{}^N(r)$ will be

$$L_0{}^N(r) \sim r^{-1} \cos(kr - \tfrac{1}{2}\pi - a \log 2kr + P + \sigma_0),$$

where P is some function of E that depends on the precise form of $V(r)$, and which becomes zero outside the resonance range. The wave function corresponding to (3) that describes the scattering will be

EFFECT OF RESONANCE LEVELS   141

$$\psi^N = A_0^N L_0^N(r) + \sum_{n=2}^{\infty} A_n^c L_n^c P_n(\cos\theta), \quad (7)$$

where A_0^N is a constant which must be chosen so that ψ^N represents an incident wave and a scattered wave. From (7) we obtain

$$\psi^N = \psi^c - A_0^c L_0^c + A_0^N L_0^N.$$

Since ψ^c denotes an incident wave and a scattered wave, we must choose A_0^N so that

$$A_0^N L_0^N - A_0^c L_0^c$$

represents a scattered wave only. This gives

$$A_0^N = e^{\,P + i\sigma_0}/k,$$

and for the scattered wave

$$\left\{ f^c(\theta) + \frac{1}{2ik}[e^{2iP} - 1] \right\} S. \quad (8)$$

The scattered intensity is given by the square of the modulus of this, and is most conveniently expressed as the ratio of the scattered intensity to the intensity given by the Rutherford formula; we obtain for this ratio

$$\left\{ \cos(\alpha \log \sin^2 \tfrac{1}{2}\theta) - \frac{1}{\alpha} \sin^2 \tfrac{1}{2}\theta \sin 2P \right\}^2$$

$$+ \left\{ \sin(\alpha \log \sin^2 \tfrac{1}{2}\theta) + \frac{1}{\alpha} \sin^2 \tfrac{1}{2}\theta (1 - \cos 2P) \right\}^2$$

where

$$\alpha = \frac{2\pi \cdot 2Z\epsilon^2}{hV} = \frac{2Z}{137}\frac{c}{v},$$

For scattering[†] at 180°, this takes the simple form

$$\left\{ 1 - \frac{1}{\alpha} \sin 2P \right\}^2 + \left\{ \frac{1}{\alpha}(1 - \cos 2P) \right\}^2.$$

[†] The recoil of the nucleus is neglected.

Although it is not possible, for a given value of the energy E of the incident α-particle, to calculate the phase P unless we know the exact form of $V(r)$, it is easy to see that as E passes over the range in which there is resonance, P increases through an amount equal to π. This may be seen as follows. Let f_1, f_2 be two solutions of (4) (for the case $n = 0$) in the region where $V > E$. Let f_1 be the solution that *decreases* as r increases, and f_2 a solution that increases exponentially; both solutions are taken everywhere positive in the region considered. Such solutions clearly exist.

Now let us integrate (4) outwards. It we take E equal to E_0, then in the range $V > E$, the solution will be Af_1. For any other value of E, it will be of the form

$$A(f_1 + af_2). \tag{9}$$

It is clear that as E passes through the value E_0, a changes sign; it is further clear that for $E < E_0$, a is positive and for $E > E_0$, a is negative.

Now let us consider the wave function outside the nucleus; it will have the form of a sine curve, and to find the phase, we have to fit it on to the solution (9) at the point where $E - V$ becomes positive. If E is not near to E_0 (i.e., not in the resonance range) af_2 will be very much greater than f_1. Thus the wave function for values of E slightly less than E_0, but not in the resonance range, is equal and opposite to the wave function for E slightly greater than E_0. In other words, as E increases through the resonance range, P increases from 0 to π.

We can thus form an idea of the form of the anomalous scattering curve in the neighbourhood of a resonance level, if artificial disintegration does not take place. Unfortunately, the anomalous scattering has not been investigated experimentally for any element that cannot be disintegrated, except H and He, in which cases it is not likely that there would be a resonance level. Figure 1 shows the type of curve that should occur. The scattering at 180°, divided by the inverse square law value is plotted against E. We have taken $Z = 8$ (oxygen) and $v = 1 \cdot 5 \times 10^9$ cm/sec. The breadth of the whole curve will depend on the breadth of the resonance level. We do not suggest that there actually is a virtual level in oxygen for this energy.

It might be thought that it would be incorrect to treat the scattering in this way, because we have assumed that the wave is in a steady state, and the time taken to reach a steady state is comparable with $T = \zeta^2 \tau$ (the time of decay from the level considered), and this is much greater (say 100 times) than the time taken for the α-particle to pass the nucleus. The objection is not valid. If the

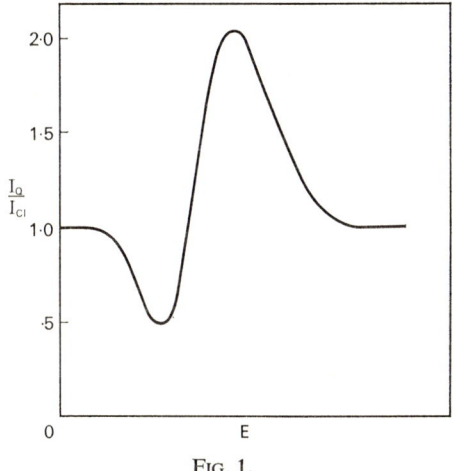

Fig. 1.

energy of the α-particle is known with an error much less than the breadth of level, so that our scattering formula can be applied, then the position of the α-particle is described by a wave packet of so great a length that the time taken for it to pass the nucleus is much greater than T, so that after a time shorter than this we may consider the wave to be in a steady state.

Section 3

We now pass on to the case when the proton can be knocked out of the nucleus. Let us denote by r, or by $r\theta\phi$, the position of

144 NUCLEAR REACTIONS

the α-particle in the nucleus, and by r' the position of the proton. Let us further denote by $\psi(r')$ the wave function that describes a proton bound in the nucleus, and by $\chi(r)$ the wave function that describes the α-particle in the stable state into which it is supposed to fall. Let $2\pi/k$, $2\pi/k'$ denote the wave-lengths of the incident α-particles and the emitted protons respectively. Then to describe the collision we want a wave function with the following asymptotic form:

$$[I + f(\theta) r^{-1} e^{ik\rho}] \psi(r') + cr'^{-1} e^{ik'\rho'}\chi(r). \tag{10}$$

Here $k\rho$ is written for $kr - i\alpha \log kr$, etc. With this wave function $|f(\theta)|^2 d\omega$ is the effective cross-section for scattering into a solid angle $d\omega$, and $4\pi \dfrac{k'}{k} |c|^2$ is the effective cross-section for disintegration. These are what we want to find.

As before, it is only particles of zero angular momentum (i.e., those described by the first term of (3)), that are affected by the resonance level. All but the first term of (3) must be multiplied by $\psi(r')$ simply; the first term must be replaced by a solution Ψ of the (unknown) wave equation for α-particle and proton; for large r or r' this solution must have the form

$$\Psi(r, r') \sim D [r^{-1} (b e^{ik\rho} + e^{-ik\rho})\psi(r') + dr'^{-1}e^{ik'\rho'}\chi(r)], \tag{11}$$

b and d depend on the wave equation; D is arbitrary and must be chosen as follows. In order that (10) shall be of the required form,

$$\Psi(r, r') - L_0^c(r)\psi(r')$$

must represent outgoing waves only. That means that it must contain no term of the type $r^{-1}e^{-ik\rho}$. This determines D. We obtain

$$D = - A_0^c e^{-i\sigma_0}/2i = - 1/2ik. \tag{12}$$

Further, for the cross-section for capture we have, since by (10) and (11), c is equal to Dd

$$a(E) = 4\pi \frac{k'}{k} | Dd |^2.$$

Now (11) is a solution of the wave equation for the α-particle and the proton in the field of the nucleus, although it is not the particular solution that describes the physical process, the incident α-particle not being represented by a plane wave. (11) describes a spherical wave of α-particles converging on the nucleus, a reflected wave, and an outgoing proton wave. Since (11) is a solution of the wave equation, it is subject to the conservation theorem; the number of α-particles falling on the nucleus, *minus* the number of particles reflected, is equal to the number of protons emitted. That is to say,

$$4\pi \mid D \mid^2 \{1 - \mid b \mid^2\} = 4\pi \mid d \mid^2 \frac{k'}{k} \mid D \mid^2. \tag{13}$$

Now this gives us an upper limit to the number of protons that can be emitted; for the right-hand side of equation (13) is the number of protons actually emitted, or more exactly the effective cross-section that must be struck for disintegration. The left-hand side of (13) has no simple physical interpretation, since it refers to just one component of a plane wave. D is, however, known; thus we can give the right-hand side an upper limit which it reaches when $b = 0$. We see that the effective cross-section for capture, $a(E)$ cannot be greater than

$$4\pi \mid D \mid^2 = \pi/k^2 = \lambda^2/4\pi, \tag{14}$$

where λ is the wave-length of the α-particle. It is to be remembered that this result applies only when the energy of the α-particle lies within the resonance range. The cross-section for capture does *not* therefore increase indefinitely for increasing sharpness of the resonance level.

If the level is not an S level, but one with azimuthal quantum number n, (14) must be replaced by

$$a(E) \leqslant (2n + 1)\lambda^2/4\pi. \tag{15}$$

For comparison we state here that the effective area that must be hit for scattering through an angle greater than 90° is, with Coulomb forces

$$\left(\frac{2Zc}{137v}\right)^2 \frac{\lambda^2}{4\pi}.$$

We must now consider under what circumstances the quantity $a(E)$ is likely to attain, for some value of E in the neighbourhood of the resonance level, the maximum value (15). We shall obtain the result, that if the interaction energy between the proton and the α-particle is large, then, provided that energetic relations allow of a proton being expelled at all, $a(E)$ will only show a broad and weak maximum, and never rise to a value approaching (15). The condition that $a(E)$ should show a sharp maximum and approach the value (15) is that the average value of the interaction energy between the α-particle and the proton, for all positions of the α-particle in the nucleus, should be small, of the order of magnitude of E/ζ, where E is the energy of the α-particle. If the interaction is greater than this, the maximum of $a(E)$ becomes broad and weak; if it becomes less, then the maximum remains sharp, but does not rise to the value (15).

The reason for this is that the breadth of the resonance level is inversely proportional to the time that the α-particle stays in that level. If the interaction between the α-particle and the proton is large, this time will be determined more by the probability of the α-particle dropping into a stable level than by the probability of the α-particle escaping. Therefore, the breadth of the resonance level will be very much greater than E/ζ^2.

This may be proved as follows. We shall make calculations using the potential field shown by the line $SABCDER$ in Fig. 2, though our results are true for any other field of the usual type assumed for a nucleus. It will simplify our argument if we assume that $\psi(r')$, the wave function for a proton bound in the nucleus, vanishes outside a sphere of radius OA, OA being less than OB in Fig. 2. We shall also assume that for distances greater than OA the forces acting on the proton are negligible. Both these assumptions could be dispensed with, but their introduction shortens the proof. Consider the function $\Psi(r, r')$ (equation (11)) that describes an incident spherical wave falling on the nucleus, a reflected wave and an outgoing proton wave. If r or r' lies in the region AB, then Ψ will have the form

$$G[r^{-1}(\beta\,e^{ikr} + e^{-ikr})\psi(r') + r'^{-1}\,\delta\,e^{ik'r'}\,\chi(r)].$$

G must be so chosen so that for large r, outside the potential barrier, the amplitude of the incident wave is $-1/2ik$—cf. equation (12). G therefore depends on the height of the potential barrier, i.e., on ζ. β and δ, however, do not depend on the height of the potential barrier, but only on the interaction between the α-particle and the

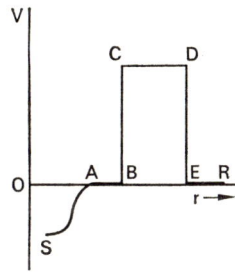

Fig. 2.

proton. They are connected by an equation similar to (13), namely,

$$1 - |\beta|^2 = \frac{k'}{k}\,|\delta|^2.$$

If the interaction is small, $|\delta|$ is small; if it is large, $|\delta|$ is of order of magnitude unity. $a(E)$ is equal to

$$4\pi k'/k\,|\,G\delta\,|^2.$$

We shall now show that unless δ is small, G must be small, and so $a(E)$ is much less than (15); but that for a certain optimum small value of δ, $a(E)$ may approach this value.

Let us denote the length OB by R_1 and the length OE by R_2. The height BC of the potential wall will be denoted by K.

To find G we have only to write down the forms of the α-particle wave in $\Psi(r, r')$ in the three regions AB, BE and ER, and put in the boundary conditions, that the function and its differential coefficient must be continuous at B and E. We have for this wave,

$$r^{-1} G(\beta e^{ikr} + e^{-ikr}) \qquad r < R_1,$$
$$r^{-1} (pe^{\gamma r} + qe^{-\gamma r}) \qquad R_1 < r < R_2$$
$$- r^{-1}/2ik(be^{ikr} + e^{-ikr}) \qquad R_2 < r.$$

Here

$$\gamma^2 = 8\pi^2 m(K - E)/h^2.$$

p, q are constants; for b see equation (11).

Solving for G as a function of k, γ and β we obtain

$$|G| = \frac{2\zeta/\gamma}{(1 + k^2/\gamma^2) |\zeta^2 (\beta' + e^{-i\alpha}) - (\beta' + e^{i\alpha})|}$$

where

$$e^{i\alpha} = (1 + ik/\gamma)/(1 - ik/\gamma), \qquad \beta' = \beta e^{2ikR_1},$$

and ζ denotes $e^{\gamma(R_2 - R_1)}$, the (large) constant that determines the time taken for an α-particle to escape from the nucleus.

It will be seen that G is small, of the order of magnitude of ζ^{-1}, unless $|\beta' + e^{-i\alpha}|$ is small. But this implies that $|\beta|$ is nearly equal to unity, and thus that $|\delta|$ is much less than unity. Thus resonance can only occur if the interaction between the α-particle and the proton is small.

Let us consider the case when $|\beta|$ is nearly equal to unity, so that resonance can occur. We put β' equal to $(1 - \epsilon) e^{i\pi - i\alpha}$, ϵ being real and small. Then to a sufficient approximation we have for $|G|$,

$$|G| = \frac{2\zeta/\gamma}{1 + k^2/\zeta^2} |\epsilon \zeta^2 + 1 - e^{2i\alpha}|^{-1}$$

and for $|\delta|$,

$$\frac{k'}{k} |\delta|^2 = 2\epsilon.$$

For $a(E)$ we have, therefore,

$$a(E) = \frac{8\lambda^2}{\pi} \left[\frac{k/\gamma}{1 + k^2/\gamma^2}\right]^2 \frac{\epsilon\zeta^2}{(\epsilon\zeta^2 + 1 - \cos 2\alpha)^2 + \sin^2 2\alpha}.$$

It is clear that if $\epsilon \sim \zeta^{-2}$ then this function is of order of magnitude λ^2; otherwise it is much smaller. Thus for resonance we must have δ of order of magnitude ζ^{-1}.

If this theory of the resonance levels is correct, it is difficult to reconcile the results of Pose, who finds quite sharp resonance levels in Al, with the results of experiments on α-particles of sufficient energy to pass over the top of the potential wall. These latter experiments show that $a(E)$ is of the order of magnitude of the cross-section of the nucleus. Thus the interaction between the α-particle and proton must be large. To explain the occurrence of resonance levels, on the other hand, one must assume that for rather lower energies of the α-particle, this interaction is small. This seems difficult to explain.

We shall now investigate the anomalous scattering of α-particles for energies near a resonance level, assuming that all the α-particles that go in are captured, so that $a(E)$ rises to its maximum value. Actually this may not be the case. For the amplitude of the scattered wave in (10) we have, if the level is of angular momentum n,

$$f(\theta) = f^c(\theta) - (2n + 1) e^{2i\sigma_n}/2ik.$$

We have, therefore, that the ratio of the number scattered to the number given by the inverse square law formula should *rise* to the value

$$1 + (2n + 1)^2/\alpha^2, \qquad \alpha = \frac{2Z}{137}\frac{c}{v}.$$

This formula refers to scattering at 180°. The scattering at other angles may be found in the same way.

Section 4. Comparison with Experiment

From the number of protons (10 per 10^8) observed by Pose, and our expression for the maximum value possible for $a(E)$, we can

150 NUCLEAR REACTIONS

form an estimate of the minimum breadth that the levels in aluminium can have.

We first require an expression for the number of protons produced per million α-particles. Let $a(E)$ be the effective cross-section that an α-particle of energy E must hit in order to eject a proton. Then the probability that an α-particle will eject a proton, while travelling a distance dx through a substance containing n atoms per unit volume is

$$na(E)dx.$$

The probability p therefore that it will produce a proton during the whole length of its path is

$$p = n \int_{v_0}^{0} a(E) \frac{dX}{dv} dv,$$

v_0 being the initial velocity of the α-particle. Now if x_0 be the total distance that the α-particle travels, we have, by the Thomson–Whiddington law

$$v^3 = k(x_0 - x),$$

and therefore

$$p = \frac{n}{k} \int_0^{v_0} a(E) 3v^2 \, dv. \tag{16}$$

Now n/k is independent of the number of atoms per unit volume, and is proportional to the atomic stopping power of the substance in question. For Al we have

$$n/k = 3 \cdot 35 \times 10^{-8}.$$

From (16), therefore, we obtain the number of protons per particle. Pose finds about 10 protons per 10^8 α-particles. (16) must, therefore, be equated to 10^{-7}. We obtain

$$\int_0^{v_0} a(E) v^2 \, dv \sim 0 \cdot 9. \tag{17}$$

Now if we assume that $a(E)$ is equal to $\lambda^2/4\pi$ in a range of velocities Δv and zero elsewhere, (17) may be replaced by $v^2 \Delta v \lambda^2/4\pi$. If we take the velocity corresponding to the resonance level at $3\cdot 9 \times 10^6$ volts ($v = 1\cdot 37 \times 10^9$ cm/sec), then for $a(E)$ in the resonance level we find

$$a(E) = 4\cdot 15 \times 10^{-26} \text{ (cm)}^2$$
$$= \pi(1\cdot 15 \times 10^{-13} \text{ cm})^2. \qquad (18)$$

For $\Delta v/v$ we find $0\cdot 03$, and therefore

$$\Delta E/E \sim 0\cdot 06$$

and

$$\Delta E \sim 2 \times 10^5 \text{ electron volts.} \qquad (19)$$

The ejected protons, therefore, should have energies distributed in a range of at least this amount. This is 2 or 3 times as much as is likely from Pose's experiments; however, the value (18) for $a(E)$ is calculated on the assumption that level is an S level. If it were a P level then $a(E)$ would be three times this, and ΔE therefore one-third of the value given by (19).

In Blackett's experiments on the artificial disintegration the cross-section for capture, averaged over the whole path of the α-particle, is about

$$\pi(2 \times 10^{-13} \text{ cm})^2.$$

This is about one-third of the total cross-section of the nucleus. Of course, in this case more than one term in the expansion of the incident wave will be "responsible" for the disintegration. A rough calculation shows that an α-particle with impact parameter 3×10^{-13} cm has four "quanta of angular momentum". We should expect, therefore, that the disintegration is due to about four terms of the series (3).

3

Neutron Capture and Nuclear Constitution†

NIELS BOHR

AMONG the properties of atomic nuclei disclosed by the fundamental researches of Lord Rutherford and his followers on artificial nuclear transmutations, one of the most striking is the extraordinary tendency of such nuclei to react with each other as soon as direct contact is established. In fact, almost any type of nuclear reactions consistent with energy conservation seems likely to occur in close nuclear collisions. In collisions between charged particles and nuclei, contact is, of course, often prevented or made less probable by the mutual electric repulsion; and the typical features of nuclear reactions are therefore perhaps most clearly shown by neutron impacts. Already in his original work on the properties of high-speed neutrons Chadwick recognized their great effectiveness in producing nuclear transmutations.[1] Especially after the discovery of artificial radioactivity by Mme and M. Joliot-Curie, most instructive evidence regarding nuclear reactions has been obtained through the researches of Fermi and his collaborators on radioactivity produced by bombardment with high-speed neutrons as well as with neutrons of thermal velocities.[2]

A typical result of the experiments with high-speed neutrons is the great probability that a collision with a nucleus of not too large atomic number will give rise to the ejection of an α-ray or a proton, accompanied by the capture of the neutron and the formation of a nucleus of a new element which in general will possess

† *Nature* **137**, 344 (1936). Address delivered on January 27 before the Copenhagen Academy (Kgl. Danske Vidensk. Selskab.).

NEUTRON CAPTURE AND NUCLEAR CONSTITUTION 153

β-ray radioactivity. The effective nuclear cross-sections for collisions with such effects are in fact of the same order of magnitude as the cross-sections responsible for simple scattering of high-speed neutrons by nuclei, which in turn agree with ordinary estimates of nuclear dimensions. Another typical experimental result is the surprisingly great tendency even for a fast neutron in collision with a heavy atom to attach itself to the nucleus with the emission of γ-radiation and the formation of a new isotope which may be stable or radioactive according to the circumstances. In fact, for processes of this kind cross-sections are found which although several times smaller are still of the same order of magnitude as nuclear dimensions.

Capture processes of high-speed neutrons of the last mentioned type are especially significant in offering a direct course of information about the mechanism of collision between the neutron and the nucleus. Indeed, the remarkable sharpness of the lines of the characteristic γ-ray spectra of radioactive elements proves that the lifetime of the excited nuclear states involved in the emission of such spectra is very much longer than the periods, *circa* 10^{-20} sec, of these lines themselves. In order that the probability of emission of a similar radiation during a collision between a high-speed neutron and a nucleus shall be large enough to account for the experimental cross-sections for these capture processes, it is therefore clear that the duration of the encounter must be extremely long compared with the time interval, *circa* 10^{-21} sec, which the neutron would use in simply passing through a space region of nuclear dimensions.

The phenomena of neutron capture thus force us to assume that a collision between a high-speed neutron and a heavy nucleus will in the first place result in the formation of a compound system of remarkable stability. The possible later breaking up of this intermediate system by the ejection of a material particle, or its passing with emission of radiation to a final stable state, must in fact be considered as separate competing processes which have no immediate connexion with the first stage of the encounter. We have here to do with an essential difference previously not clearly recognized between proper nuclear reactions and ordinary collisions among fast particles and atomic systems, which have been our main source of information about the structure of the atom. In fact, the possibility of counting by means of such collisions the individual atomic particles and of studying their properties is due above all to the openness of the systems concerned, which makes an energy exchange between the separate constituent particles during the encounter very unlikely. In view of the close packing of the particles in nuclei we must be prepared, however, for just such energy changes to play a predominant part in typical nuclear reactions.

If, for example, we consider an encounter between a high-speed neutron and a nucleus, it is obviously not permissible to compare the process to a simple deflection of the path of the neutron in the inner nuclear field, possibly combined with a collision with a separate nuclear particle, resulting in the ejection of the latter. On the contrary, we must realize that the excess energy of the incident neutron will be rapidly divided among all the nuclear particles with the result that for some time afterwards no single particle will possess sufficient kinetic energy to leave the nucleus. The possible subsequent liberation of a proton or an α-particle or even the escape of a neutron from the intermediate compound system will therefore imply a complicated process in which the energy happens to be again concentrated on some particle at the surface of the nucleus.

At the moment it is scarcely possible to form a detailed picture of such processes. In fact, we must recognize that we have no justification even for assuming the existence within nuclei of the particles set free in nuclear disintegrations. In particular, the well-known difficulties of attributing within a space region of nuclear dimensions an individual existence to charged particles with so small a rest mass as have electrons and positrons, forces us to consider β-ray disintegration as a process by which an electron is created as an entity in a mechanical sense.[3] In this respect the situation is, of course, essentially different for the heavier particles emitted in nuclear disintegrations, like neutrons, protons and α-rays. Especially the fact that all nuclear masses in the first approximation are integral multiples of a unit nearly equal to the neutron mass, makes it very reasonable to regard particles of such masses as mechanical entities within nuclei. On account of the small difference between the masses of the neutron and the proton compared with the binding energies in nuclei measured by their so-called mass defect, it would, however, seem more hypothetical to assume the existence in nuclei of particles with the same electric and magnetic properties as those possessed by free neutrons and protons. In view of the scarcity of our knowledge of the extraordinary dense state of matter with which we have to do in nuclei, we may rather regard the integral values of unit electric charge possessed by nuclei and their disintegration products as a fundamental aspect of the atomicity of electrification, which cannot be accounted for by present theories of atomic constitution.

Quite apart from the problem of the nature of the nuclear constituents themselves, which is not of direct importance for the present discussion, it is, at any rate, clear that the nuclear models hitherto treated in detail are unsuited to account for the typical properties of nuclei for which, as we have seen, energy exchanges between the individual nuclear particles is a decisive factor. In fact, in these models it is, for the sake of simplicity, assumed that the state of motion of each particle in the nucleus can, in the first approximation, be treated as taking place in a conservative field of force, and can therefore be characterized by quantum numbers in a similar way to the motion of an electron in an ordinary atom. In the atom and in the nucleus we have indeed to do with two extreme cases of mechanical many-body problems for which a procedure of approximation resting on a combination of one-body problems, so effective in the former case, loses any validity in the latter where we, from the very beginning, have to do with essential collective aspects of the interplay between the constituent particles.

In this connexion it is of importance to remember also that the successful quantum mechanical explanation of the simple law combining the life-time of α-ray products with the energy of the emitted particles, is essentially independent of any special assumption regarding the behaviour of the individual particles in the nucleus. In fact, on account of the extremely long lifetime of these products compared with all proper nuclear periods, the probability of such disintegration depends in the first approximation only upon the electric field outside the nucleus, which constitutes a so-called potential barrier hindering the escape of the α-rays. It is even very doubtful that α-particles exist in nuclei in the manner assumed in present theories of α-ray decay. Indeed, the frequent appearance of α-rays as a result of natural and artificial nuclear disintegrations may rather be explained by the fact that energy is set free by the very formation of α-particles, and that the liberation of such particles might thus involve a smaller degree of concentration of excess energy than the liberation of protons or neutrons. So far, the study of the α-ray disintegrations and their intimate connexion with the γ-ray spectra, especially cleared up by Gamow, gives us, therefore, information only about the possible values of the energy and to a certain extent of the spin momenta for the stationary states of the nuclear systems concerned.

The circumstance that the nuclear states involved in the last mentioned phenomena are found to represent a discrete distribution of very sharp energy levels might perhaps at first sight seem to contrast with our assumptions of the existence of a semi-stable intermediate state of the compound system formed by neutron collisions within an apparently continuous range of the kinetic energy of the incident neutron. We must realize, however, that in the impacts of high-speed neutrons we have to do with an excitation of the compound system far greater than the excitation of ordinary γ-ray levels. While the latter at most amounts to a few million volts, the excitation in the former case will considerably exceed the energy necessary for the complete removal of a neutron from the normal state of the nucleus. The measurements by Aston of the mass differences of isotopes show that this energy is about ten million volts.

This striking difference in the level schemes for low and high excitations of heavy nuclei, is, however, just what we would expect according to the view of nuclear reactions here discussed. In contrast to the usual view, where the excitation is attributed to an elevated quantum state of an individual particle in the nucleus, we must in fact assume that the excitation will correspond to some quantized collective type of motion of all the nuclear particles. On account of the rapid increase of the possibilities of combination of the proper frequencies of such motions for increasing values of the total energy of the nucleus, we should therefore expect that the distance between neighbouring levels would become very much smaller for the high excitation concerned in neutron collisions than in the ordinary γ-ray levels where we have probably to do with states of collective motions of the most simple types. Even for excitations where the levels are very close together the probability of radiative transition will not, however, on this view be very much different from that in the lower γ-ray states and any material increase in the width of the levels will not arise, until the probability of escape of material particles becomes comparable with the radiation probability.

Now, in experiments on high-speed neutron impacts on heavy nuclei, the effective cross-section for scattering is normally several times larger than the cross-section for capture. Accordingly, we must conclude that in this case the probability of the escape of a neutron from the compound system is greater than the probability of radiative transitions and that the energy levels of the semistable state are therefore somewhat broader than ordinary γ-ray levels. This circumstance, together with the rapidly decreasing distance between neighbouring levels in the energy region concerned, makes it indeed very likely that such levels will not here be separated at all, as is required for the explanation of the apparently non-selective character of the capture phenomena. For decreasing velocities of the incident neutrons, however, escape of a neutron from the compound system will rapidly become very improbable, corresponding to the decreasing probability of the necessary concentration of the excess energy of the system on a particular neutron. The sharpness of the levels of the intermediate state must therefore be expected to approach that of the γ-ray levels, as soon as the kinetic energy of the free neutrons becomes small compared with the total excitation energy in this state.

158 NUCLEAR REACTIONS

Most interesting support for these considerations is afforded by the remarkable phenomena of selective capture of neutrons of very small velocities. Working with neutrons of temperature velocity obtained by passing neutron beams through thick sheets of substances containing hydrogen, Fermi and his collaborators found, as is well known, values for the effective cross-sections for neutron capture, which vary in a most capricious way from element to element. While for most elements these values were of the same order of magnitude or not much larger than ordinary nuclear cross-sections, values several thousand times larger were found in certain irregularly distributed elements or isotopes. These at first sight most surprising effects must obviously be attributed to the fact that for such slow neutrons the de Broglie wave-length is very large compared with nuclear dimensions and that, therefore, the simple ideas of path and collision, which can be applied at any rate approximately to high-speed neutron impacts, here fail completely.

Instructive attempts have also been made to explain the phenomenon of selective capture as a quantum mechanical resonance phenomenon, due to the close coincidence between the energy of some almost stable stationary states of the neutron within the nucleus and the sum of the energies of the initial state of the nucleus and of the free neutron.[4] These theories, in which the state of motion of the neutron within the nucleus is treated as that of a particle in a conservative field of force, have failed, however, to account for the fact that the cross-section for neutron scattering in all selective absorbing elements investigated is much smaller than the cross-section for capture. It is time that the large probability of reflection of the waves describing the behaviour of the neutron in the nucleus—arising from the fact that its wave-length here is very short compared with the wave-length for the free motion of the neutron—implies that the mean time interval which a neutron may be said to stay in a nucleus is very long compared with the time interval a high-speed neutron on such a model would take in passing through the nucleus. Still, even in the case of complete resonance, the probability of neutron escape is in this way found to be larger than the probability for emission of radiation. From the far more intimate interaction between the neutron and the nucleus, which the explanation of high-speed neutron capture demands, this remarkable absence of selective scattering of very slow neutrons is, however, just what we should expect for small excess energy, on account of the vanishing probability of neutron escape compared with that of radiative transition.

Moreover experiments of Fermi and others[5] during the last few months have revealed an extreme sensitiveness of the phenomena of selective neutron capture for small variations in the neutron velocity which necessitates a degree of resonance quite incompatible with the above-mentioned nuclear model. In fact, by the filtration of low-speed neutron beams through thin sheets of different selective absorbing elements great modifications in the cross-sections of selective capture were obtained, showing that the resonance is restricted to narrow neutron energy regions which are differently placed for different selective absorbers. By using for comparison the capture of neutrons in light elements resulting in the ejection of α-particles, where the selectivity is much less pronounced—and where therefore from general quantum mechanical arguments the probability of capture within the energy region concerned must be expected to be inversely proportional to the neutron velocity—it has even been possible to conclude that the energy region of resonance for certain selective absorbing elements is confined within a fraction of a volt.[6]

From this small breadth of the energy levels of the compound system formed by low-speed neutron capture, we arrive by a simple statistical consideration of the occurrence of selective capture among the heavier elements at an estimate of about ten volts for the distance between neighbouring energy levels for the excitation concerned in these phenomena. This is not only in full agreement with the conclusions about the close distribution of energy levels of highly excited nuclei to which we were led through the discussion of the non-selective capture of the high-speed neutrons; but the extreme sharpness of the levels with which we are concerned in the phenomena of selective neutron capture offers also most interesting support for our primary assumption of the long lifetime of the intermediate state in neutron collisions. In fact, the narrowness of the levels of the compound system proves in a striking way the extreme smallness of the probability of radiative transitions in nuclei and leads to an estimate for the duration of an encounter between a high-speed neutron and a nucleus as large as a million times the interval which the neutron would use in simply traversing the nucleus.

The lack of selectivity in high-speed neutron impacts concerns strictly speaking only the probability of neutron capture by the nucleus and the ejection of a material particle from it. The detailed course of these phenomena will, however, depend in general essentially on the level system of the nucleus finally formed. In fact after the collision process this system must be in some stationary state and if the kinetic energy of the incident neutron is not very large the states between which there can be a choice will all be in the region of ordinary discrete γ-ray levels. If then the kinetic energy of the neutrons impinging on a heavy nucleus is smaller than the lowest excited level of this nucleus, any neutron escaping from the compound system will necessarily possess the same energy as the incident neutron. In the case, however, of neutron impacts with higher energy there is obviously a certain probability that the nucleus may be left in an excited state after the escape of a neutron with a correspondingly smaller energy. Actually, the probability of the process following such a course, which implies a smaller concentration of the excess energy of the compound system on the escaping neutron, may often be considerably greater than the probability of neutron escape without excitation. There seems, too, to be experimental evidence for the occurrence of nuclear excitation in neutron collisions, namely in the observation of energy loss of high-speed neutrons traversing substances of high atomic weight,[7] where a direct transfer of translational energy between the neutrons and the nuclei would be expected to be negligibly small.

As was mentioned earlier, collisions between high-speed neutrons and the nuclei of elements of small atomic number will in most cases result in the ejection of an α-ray or a proton. We may conclude here also from the great cross-sections for collision of such effects, that the encounter leads in the first place to the formation of a semi-stable compound system with a continuous range of energy levels. Even though the lifetime of this system may be very much shorter than that of the γ-ray states of heavy nuclei, we must still realize that the subsequent escape of α-rays or protons necessitates a separate concentration process for the excess energy and that in particular we cannot draw any decisive conclusion from these phenomena about the presence of such particles in nuclei under normal conditions. For example, the great probability of emission of α-rays compared with neutron escape from the compound system must, as already indicated, rather be explained by the comparatively small degree of energy concentration involved in the former process. As regards the emission of charged particles we must of course also take into account the repulsion from the rest of the nucleus and in particular the greater difficulty experienced by a charged than by an uncharged particle with the same final kinetic energy in passing the potential barrier round the nucleus.

As has often been pointed out, the last circumstance offers a simple explanation not only of the rapid fall in the output of α-particles and protons as a result of high-speed neutron impacts for increasing nuclear charge, but also of the decrease with increasing neutron energy of the ratio between the probabilities of ejection of these two differently charged kinds of particles. The probability of the nucleus being left after the ejection of such particles in the normal or in an excited state depends in each case on the distribution of the energy levels of the final system—which are in general more separated for light than for heavy nuclei—and also on the balance between, on one hand, the greater facility of faster particles than of slower ones in penetrating the potential barrier and, on the other hand, the greater demands for energy concentration in the former than in the latter case. Similar considerations will apply for the finer details of the ordinary α-ray disintegrations like the weak groups of long range α-particles and the fine structure of the stronger α-ray lines.

In the case also of nuclear transmutation caused by the impact of charged particles as well as for the nuclear disintegration produced by γ-rays, the formation of an intermediate semi-stable compound system seems decisive for the explanation of the great variety of the phenomena. Besides typical non-selective effects like the ejection of neutrons and protons by fast α-rays, we meet, as is well known, with pronounced resonance effects for slower α-ray impacts, as well as in the capture phenomena of artificially accelerated protons in light nuclei. On account of the very much shorter lifetime of the intermediate state in such cases the degree of resonance here obtained is, however, much smaller than for selective neutron capture by heavy nuclei. In this connexion it is perhaps not out of place to note that expressions like α-ray levels or proton levels, such as are used in the ordinary discussion of these effects, based on the attribution of the excitation to a single nuclear particle, lose all meaning on the view of nuclear excitation adopted here. In fact, the essential feature of nuclear reactions, whether incited by collision or by radiation, may be said to be a free competition between all the different possible processes of liberation of material particles and of radiative transitions, which can take place from the semi-stable intermediate state of the compound system.

A detailed discussion from this point of view of the available empirical evidence regarding spontaneous and induced nuclear transmutations will be published shortly[8] in collaboration with Mr. F. Kalckar, who has given me most valuable assistance in tracing the consequences of the general argument here developed. There we shall also discuss the limitation of this argument in the case of very light nuclei like the deuteron, where the distinction between the mechanism of the storing of the energy in the nucleus and the mechanism of the liberation of particles, so pronounced for the reactions of heavy nuclei, gradually loses its significance. Here, however, I should still like briefly to indicate what modifications in the preceding considerations are to be expected even for heavy nuclei should the energy of the intermediate system too far exceed that of its normal state. Even if we could experiment with neutrons or protons of energies of more than a hundred million volts, we should still expect that the excess energy of such particles, when they penetrate into a nucleus of not too small mass, would in the first place be divided among the nuclear particles with the result that a liberation of any of these would necessitate a subsequent energy concentration. Instead of the ordinary course of nuclear reactions we may, however, in such cases expect that in general not one but several charged or uncharged particles will eventually leave the nucleus as a result of the encounter. For still more violent impacts, with particles of energies of about a thousand million volts, we must even be prepared for the collision to lead to an explosion of the whole nucleus. Not only are such energies, of course, at present far beyond the reach of experiments, but it does not need to be stressed that such effects would scarcely bring us any nearer to the solution of the much discussed problem of releasing the nuclear energy for practical purposes. Indeed, the more our knowledge of nuclear reactions advances the remoter this goal seems to become.

In concluding this address, I should just like to point out that even if the problem of nuclear constitution does lack the special simplicity in a mechanical respect characteristic of the structure of the atom which has so much facilitated the disentanglement of the relationships of the elements as regards their ordinary physical and chemical properties, it presents, nevertheless, as I have tried to show, peculiar facilities for a comprehensive interpretation of the characteristic properties of nuclei in allowing a division of nuclear reactions into well separated stages to an extent which has no simple parallel in the mechanical behaviour of atoms.

References

[1] J. Chadwick, *Proc. Roy. Soc.* **A142**, 1 (1933).

[2] E. Fermi and others, *Proc. Roy. Soc.* **A146**, 483 (1934); **149**, 522 (1935).

[3] Cf. N. Bohr, Faraday Lecture, *J. Chem. Soc.* 349 (1932), and W. Heisenberg, *Zeeman Verhandelingen*, p. 108.

[4] E. Fermi and others, *Proc. Roy. Soc.* **A149**, 522 (1935). Perrin and Elsasser, *J. Phys.* **6**, 195 (1935). Béthe, *Phys. Rev.* **47**, 747 (1935).

[5] E. Fermi and Amaldi, *La Ricercio Scientifica* **A6**, 544 (1935). Szilard, *Nature*, **136**, 849 (1935). R. Frisch, Hevesy and McKay, *Nature*, **137**, 149 (1936).
[6] R. Frisch and G. Placzek, *Nature*, **137**, 357 (1936).
[7] W. Ehrenberg, *Nature*, **136**, 870 (1935).
[8] N. Bohr and F. Kalckar, *Kgl. Dan. Vid. Selsk. Math-fys. Medd.* (in preparation).

4

Capture of Slow Neutrons[†]

G. Breit and E. Wigner

> **Abstract.** Current theories of the large cross-sections of slow neutrons are contradicted by frequent absence of strong scattering in good absorbers as well as the existence of resonance bands. These facts can be accounted for by supposing that in addition to the usual effect there exist transitions to virtual excitation states of the nucleus in which not only the captured neutron but, in addition to this, one of the particles of the original nucleus is in an excited state. Radiation damping due to the emission of γ-rays broadens the resonance and reduces scattering in comparison with absorption by a large factor. Interaction with the nucleus is most probable through the s part of the incident wave. The higher the resonance region, the smaller will be the absorption. For a resonance region at 50 volts the cross-section at resonance may be as high as 10^{-19} cm^2 and $0\cdot 5 \times 10^{-20}$ cm^2 at thermal energy. The estimated probability of having a nuclear level in the low energy region is sufficiently high to make the explanation reasonable. Temperature effects and absorption of filtered radiation point to the existence of bands which fit in with the present theory.

1. Introduction

Bethe,[1] Fermi,[2] Perrin and Elsasser,[3] Beck and Horsley[4] gave theories of the anomalously large cross-sections of nuclei for the capture of slow neutrons. These theories are essentially alike and explain the anomalously large capture cross-sections as a sort of resonance of the s states of the incident particle. Resonance is usually helpful in causing a large scattering as well as a large probability of capture and it has been shown [H.B., eq. (35)] that large scattering is to be expected by nuclei showing anomalously large capture at thermal energies. This consequence of the current

[†] *Phys. Rev.* **49**, 519 (1936).

theories is apparently in contradiction with experiment, there being no evidence of a large scattering in good absorbers. It also follows from current theories that with very few exceptions the capture cross-section should vary inversely as the velocity of the slow neutrons. Experiments on selective absorption recently performed[5] indicate that there are absorption bands characteristic of different nuclei and it appears from the experiments of Szilard[6] that these bands have fairly well-defined edges. It has been pointed out by Van Vleck[7] that it is hard and probably impossible to reconcile the difference in internal phase required by the Bethe–Fermi theory with reasonable pictures of the structure of the nucleus. The combined evidence of experimental results and theoretical expectation is thus against a literal acceptance of the current theories and it is our purpose to outline an extension which is capable of explaining the above facts by a mechanism similar to that used for the inverse of the Auger effect by Polanyi and Wigner.[8]

It will be supposed that there exist quasi-stationary (virtual) energy levels of the system nucleus+neutron which happen to fall in the region of thermal energies as well as somewhat above that region. The incident neutron will be supposed to pass from its incident state into the quasi-stationary level. The excited system formed by the nucleus and neutron will then jump into a lower level through the emission of γ-radiation or perhaps at times in some other fashion. The presence of the quasi-stationary level, Q, will also affect scattering because the neutron can be returned to its free condition during the mean life of Q. If the probability of γ-ray emission from Q were negligible there would be in fact strong scattering at the resonance, the scattering cross-section being then of the order of the square of the wave-length. Estimates of order of magnitude show that it is reasonable to assign 12 volts to the "half-value breadth" of Q due to radiation damping and that the "half-value breadth" due to passing back into the free state is about one-fortieth of the above amount. This means that when the system passes into the state Q it radiates practically immediately and the neutron has no time to be rescattered. It will, in fact, be seen from the calculations that follow that the ratio of scattering to absorption is essentially the ratio of the corresponding half-value breadths. The hardness of the emitted γ-rays is of primary importance for the small ratio of scattering to absorption because it makes the probability of γ-ray emission sufficiently high. Inasmuch as the interesting phenomena occur for low energies we may suppose that in most cases the coupling of the incident state occurs through its s state, i.e., in virtue of head-on collisions. It will be seen, however, that the possibility of obtaining observable effects by means of p states is not excluded even though it is less probable and leads to smaller cross-sections. Calculation shows that with resonances of the type considered here one may obtain appreciable probability of capture at energies of the order of 1000 volts. It is possible to have at such energies cross-sections of roughly 10^{-22} cm^2 with a half-value breadth of about 20 volts. It is therefore not necessary to ascribe all large cross-sections to neutrons of thermal velocities and the probability of finding a quasi-stationary level in a suitable region is not so small as to make the process improbable.

We are presenting below the theory of capture on this basis in some detail not because we believe it to be a final theory but because further development may be helped by having the preparatory structure well cemented.

2. Theory of Damping

The process of absorption from the continuum into a quasi-stationary level and a subsequent reemission of a photon is related to the phenomena of predissociation discussed by O. K. Rice[8] who made the first application of quantum mechanics to this type of process since Dirac's first approach.[9] It is essential for us to consider two continua and in this respect the present problem is more general. It resembles closely the problem of absorption of light from a level a to a level c which is strongly damped by radiation in jumps to a third level b. The absorption from a to c corresponds to the transition of the neutron into the quasi-stationary level and the jumps from c to b correspond to the emission of γ-rays in a transition to a more stable level of the nucleus. The absorption probabilities can be obtained by using the principle of detailed balance from the solution which represents emission[10] from the level c to the levels a, b or else by a direct application of the theory of absorption.[11] The usual theory as developed for either process is not accurate enough to represent the effect of the variation of matrix elements with velocity which is essential for our purpose, inasmuch as it is responsible for the existence of two regions of large absorption. The usual type of calculation will now be generalized so as to take the variation into account.

(a) Calculation of the absorption and scattering process

Let a_s denote the probability amplitudes of states in which the neutron is free and in a state s. Similarly let b_r stand for the probability amplitude of a state in which the neutron is captured and there is a photon r emitted and let c be the probability amplitude of the quasi-stationary state having energy $h\nu$. The states r, s are here considered to be discrete but very closely spaced in energy. The average spacing of the levels r, s are written $\Delta E_r = h\Delta\nu_r$, $\Delta E_s = h\Delta\nu_s$ so that the number of levels s per unit energy range is $1/\Delta E_s$. The matrix element of the interaction energy responsible for transitions from a_s to c, c to b_r will be written, respectively,

$$M_s = hA_s, \quad M_r = hB_r. \tag{1}$$

The damping constants for c due, respectively, to the possibility of emitting a_s or b_r are then [12]

$$(4\pi\tau_a)^{-1} = \Gamma_s = [\pi\overline{\mid A_s \mid^2}/\Delta\nu_s]_{\nu_s=\nu_0}; \tag{2}$$

$$(4\pi\tau_b)^{-1} = \Gamma_r = [\pi\overline{\mid B_r \mid^2}/\Delta\nu_r]_{\nu_r=\nu_0}; \quad \Gamma = \Gamma_s + \Gamma_r,$$

where τ_a, τ_b are respective mean lives of c due to emission of a_s and b_r. The quantities Γ represent one-half of the "half-value breadth" measured in frequency. In discussing line emission and absorption[10, 11] the directional averages of $|A_s|^2$ and $|B_r|^2$ can be

CAPTURE OF SLOW NEUTRONS   169

taken for any energy within the breadth of the line because the line can be usually considered to be sharp. In the present case it will be necessary to distinguish among directional averages of $|A_s|^2$ for different energies.

The states s will be thought of as plane waves modified by a central field due to the nucleus and satisfying boundary conditions at the surface of a fundamental cube of volume V. The equations satisfied by a_s, b_r, c are

$$\left(\frac{d}{2\pi i dt} + \nu_s\right)a_s = A_s c; \quad \left(\frac{d}{2\pi i dt} + \nu_r\right)b_r = B_r c, \quad \left(\frac{d}{2\pi i dt} + \nu\right)c$$
$$= \Sigma A_s^* c_s + \Sigma B_r^* b_r. \tag{3}$$

In these equations the influence of only one quasi-stationary level is taken into account and for this reason they are not quite accurate. They are sufficiently good for the present purpose because it will be supposed that different quasi-stationary levels do not fall closely together. At $t = 0$ it will be supposed that

$$a_s = \delta_{ss_0}, \quad b_r = 0, \quad c = 0, \quad (t = 0). \tag{4}$$

An approximate solution of (3) satisfying this initial condition can be obtained by forming a linear combination of

$$c = e^{-2\pi i(\nu - i\Gamma')t};$$
$$a_s = A_s[e^{-2\pi i(\nu - i\Gamma')t} - e^{-2\pi i\nu_s t}]/(\nu_s - \nu + i\Gamma'), \tag{5}$$
$$b_r = B_r[e^{-2\pi i(\nu - i\Gamma')t} - e^{-2\pi i\nu_r t}]/(\nu_r - \nu + i\Gamma'),$$

with

$$\Gamma' = [\overline{\pi|A_s|^2}/\Delta\nu_s + \overline{\pi|B_r|^2}/\Delta\nu_r]_{\text{resonance region}}, \tag{5'}$$

and

$$a_{s_0} = e^{-2\pi i(\nu_0 - i\gamma)t}; \quad c = A_{s_0}^* e^{-2\pi i(\nu_0 - i\gamma)t}/(\nu - \nu_0 - i\Gamma),$$
$$a_s = A_s A_{s_0}^*[e^{-2\pi i(\nu_0 - i\gamma)t} - e^{-2\pi i\nu_s t}]/(\nu_s - \nu_0 + i\gamma)(\nu - \nu_0 - i\Gamma),$$
$$b_r = B_r A_{s_0}^*[e^{-2\pi i(\nu_0 - i\gamma)t} - e^{-2\pi i\nu_r t}]/(\nu_r - \nu_0 + i\gamma)(\nu - \nu_0 - i\Gamma),$$
$$\Gamma = [\overline{\pi|A_s|^2}/\Delta\nu_s + \overline{\pi|B_r|^2}/\Delta\nu_r]_{\nu_s \simeq \nu_0}. \tag{6}$$

In eq. (6) $s \neq s_0$. The quantities γ and $\nu_0 - \nu_{s_0}$ are small compared with Γ; they will go to zero with increasing volume. From (3), one finds for them the equation:

$$(\nu_{s_0} - \nu_0 + i\gamma)(\nu - \nu_0 - i\Gamma) = |A_{s_0}|^2, \tag{7}$$

so that

$$\gamma = |A_{s_0}|^2 \Gamma/[(\nu - \nu_0)^2 + \Gamma^2];$$

$$\nu_0 = \nu_{s_0} + (\nu_0 - \nu)(\gamma/\Gamma). \tag{8}$$

In obtaining eq. (7) the approximations

$$\Sigma_s' |A_s|^2 \frac{1 - e^{2\pi i(\nu_0 - \nu_s - i\gamma)t}}{\nu_s - \nu_0 + i\gamma} = \pi i \overline{|A_s|^2}/\Delta\nu_s \tag{9}$$

are made. These correspond to replacing the sums by integrals and extending the range of integration from $\nu_s = -\infty$ to $\nu_s = +\infty$ and similarly for ν_r. In addition it is supposed that $\overline{|A_s|^2}$, $\overline{|B_r|^2}$ vary so slowly through the region in which the integrand is large that they may be taken outside the integral sign. These approximations are therefore, valid only if the contributions to the sums (9a), (9b) are localized in a sharp maximum. Such a maximum exists for $\nu_s \cong \nu_0$ because: (1) γ vanishes as the fundamental volume is increased and therefore one may consider $\gamma t \ll 1$ and (2) for any $\nu_s - \nu_0$ it is possible to choose t sufficiently large to make $|\nu_s - \nu_0|t \gg 1$. For such times the most important part of the integrand oscillates rapidly with ν_s. However, for $|\nu_s - \nu_0| \sim \gamma$, the values of t which satisfy $\gamma t \ll 1$ are always such that $|\nu_s - \nu_0|t \ll 1$. The integrand is thus not oscillatory for $\nu_s = \nu_0 \pm \gamma$ and the values of $\overline{|A_s|^2}$, $\overline{|B_r|^2}$ on the right side of (9) are to be understood as corresponding to $\nu_s = \nu_0$ with an uncertainty of the order γ. It can be verified by calculation that the contribution to (9) due to a finite region at a distance $|\nu_s - \nu_0| \gg \gamma$ contributes imaginary quantities decreasing exponentially with $2\pi|\nu_s - \nu_0|t$ and real quantities which contribute to a frequency shift[9] of ν. For the present this shift will be neglected. Equations (6) are thus approximate solutions which become increasingly better as t increases, provided $\gamma t \ll 1$. In our application Γ is mostly due to the radiation damping Γ_r. The directional averages of $|B_r|^2$ vary smoothly since the energy of the γ-ray is of the order of several million volts and is large compared to Γ.

The quantity Γ' which enters (5) is not determined accurately by the present method because $\overline{|A_s|^2}$ which enters in this case is some sort of average over the resonance width. This complication causes no trouble because: (a) for times $t \gg 1/4\pi\Gamma'$ the rates of emission of states a_s, b_r are, respectively, $4\pi\gamma\Gamma_s/\Gamma$, $4\pi\gamma\Gamma_r/\Gamma$ and depend[13] only on Γ and not on Γ'; (b) the largeness of Γ_r in comparison with Γ_s makes $|\Gamma' - \Gamma| \ll \Gamma$. Thus Γ' is of importance only in determining the initial transients but not the steady rate of absorption. This can be expected from the fact that the solutions (6) represent a condition in which s_0 is absorbed at the rate $4\pi\gamma$. The addition of the "emission solution" (5) is only needed to enforce the condition $c = 0$ at $t = 0$; it modifies the emission of states b_r, a_s during times comparable with the mean life of the nucleus but leaves them unchanged over longer periods very similarly to the way in which analogous transient conditions are of no importance in the absorption of monochromatic radiation by classical vibrating systems.

The total cross-section σ which corresponds to the disappearance of the incident states s_0 is given by

$$\sigma = 4\pi\gamma V/v, \tag{10}$$

where v is the neutron velocity because the modified plane waves denoted by s were normalized in the volume V and thus represent states of density $1/V$.

The number of possible plane waves in V per unit frequency range is

$$1/\Delta\nu_s = 4\pi V/v\varLambda^2, \tag{11}$$

where $\varLambda$ is the de Broglie wave-length. From (2), (10), (11) we have

$$\sigma = \gamma\varLambda^2/\Delta\nu_s = \frac{\varLambda^2}{\pi} S \frac{\Gamma_s\Gamma}{(\nu - \nu_0)^2 + \Gamma^2}. \tag{12}$$

Here the statistical factor S takes account of the fact that the state s_0 may be more or less effective in its coupling to the quasi-stationary level than the average modified plane wave in the same energy region. If the quasi-stationary level has an orbital angular momentum $L\hbar$ and if there is no spin–orbit interaction then $|A_{s_0}|^2 = (2L+1)\overline{|A_s|^2}$ because coupling to c can take place only through $1/(2L+1)$ of the total number of states. Thus.

$$S = 2L + 1 \qquad (13)$$

in these special circumstances. For s terms $S = 1$. The total cross-section

$$\sigma = \sigma_c + \sigma_s,$$

where σ_s is the cross-section due to scattering and σ_c is the cross-section due to capture. We have

$$\sigma_c = \frac{\Lambda^2}{\pi} S \frac{\Gamma_s \Gamma_r}{(\nu - \nu_0)^2 + \Gamma^2}; \quad \sigma_s = \frac{\Lambda^2}{\pi} S \frac{\Gamma_s^2}{(\nu - \nu_0)^2 + \Gamma^2}. \qquad (14)$$

The above value of σ_s corresponds to the value $\Sigma |a_s|^2$ and does not take into account the fact that there is scattering in the absence of the quasi-stationary level. If this is strong one must correct σ_s for interference of the states s with the spherical wave present in s_0. In the applications made below the scattering effect due to either cause will be small and the correction need not be considered. According to (14) the extra scattering can be expected to be of the order Γ_s/Γ_r times the capture and is quite small for small Γ_s.

It should be noted that the order of magnitude of σ_c at resonance is changed by taking into account the radiation damping. If this were neglected and if one were to calculate simply by using Einstein's emission probability for the stationary states of matter then one would obtain an incorrect value,

$$\sigma_c' = \frac{\Lambda^2}{\pi} S \frac{\Gamma_s \Gamma_r}{(\nu - \nu_0)^2 + \Gamma_s^2}. \qquad (14')$$

CAPTURE OF SLOW NEUTRONS 173

For resonance $\sigma_c/\sigma_c' = \Gamma_s^2/\Gamma^2$ and approximately $\int \sigma_c dE / \int \sigma_c' dE$ is Γ_s/Γ_r. No paradox is involved here because it is not legitimate to apply Einstein's emission probability formula to levels separated by less than their breadth due to radiation damping. Equation (14′) gives too high values to the cross-section. If $\nu - \nu_0 \gg \Gamma$ there is no difference between σ_c' and σ_c. For sufficiently large values of $\nu - \nu_0$ the discussion which led to eqs. (13), (14) will break down because Dirac's frequency shift[9] is neglected in these formulas. A more complete formal discussion including the frequency shift is given in Appendix I. The calculation shows that one should change the frequency of the quasi-stationary level ν by

$$\nu \to \nu - \int \frac{|A_s|^2}{\Delta \nu_s} \frac{d\nu_s}{\nu_s - \nu_0} - \int \frac{|B_r|^2}{\Delta \nu_r} \frac{d\nu_r}{\nu_r - \nu_0} \qquad (15)$$

where the integrations are extended over the complete range of states s, r and where the principal values of the integrals are to be taken. The last part of (15) represents the frequency shift due to electromagnetic radiation and can be incorporated in ν as a constant because ν_0 need be varied only in a range small in comparison with the frequency of the γ-ray. It is dangerous to take this shift into account on account of the well-known inconsistency of quantum electro-dynamics. The second term on the right side of eq. (15) is due to interactions between free neutron states and the quasi-stationary state. It is physically correct and it is necessary in order to bring about agreement between (14) and calculations away from resonance by means of the Einstein emission probabilities. The shift is large in the applications. Nevertheless changes in it are small in the relatively small range of values which need be considered and its effect is therefore primarily that of displacing the resonance frequency by a constant amount.

(b) RESONANCE OF ONE-BODY SYSTEMS

The above discussion cannot be applied directly to cases in which resonance consists simply in a sharp increase of the wave function of one neutron to a maximum inside the nucleus because there is no intermediate state c under

174 NUCLEAR REACTIONS

such conditions in the same sense as in the previous section. For low velocity neutrons such resonance can be sharp for states with $L \geq 1$. Formally one could try to apply the discussion already given by starting with wave functions which are solutions of the wave equation for an infinitely high barrier somewhat outside the nucleus. The difference between the actual height of the barrier and ∞ can be then treated as a perturbation essentially responsible for the matrix elements hA_s. Such a procedure leads apparently to correct results which can be verified by other methods. It is troublesome to justify it completely because the region where the infinite barrier must be erected should be such that the wave functions within are small for all energies. It is preferable to use a more direct calculation for such a case. We consider a plane wave of neutrons incident on the nucleus. Resonance takes place to the wave functions of angular momentum $L\hbar$. We surround the nucleus by a large perfectly reflecting sphere of radius R and we calculate the rate at which states of angular momentum $L\hbar$ disappear by radiation. There is no essential restriction on the possibility of forming wave packets out of the plane waves if we admit only those states L which satisfy the boundary conditions on the sphere. The radius will be made finally infinitely large and the spacing between the levels infinitely small. This provides the necessary flexibility for the formation of the wave packets.

The spacing between successive possible neutron levels is given by

$$\Delta \nu = v/2R. \tag{16}$$

The radial function will be expressed as F/r where F will be by definition a sine wave with unit amplitude at a large distance from the nucleus. The normalized wave function is then $Y_L(F/r)(2/R)^{\frac{1}{2}}$ where Y_L is a spherical harmonic normalized so as to have $\int |Y_L|^2 d\Omega = 1$. The wave function for the bound state will be written

$$Y_{L\pm 1} f/r; \int_0^\infty f^2 dr = 1. \tag{17}$$

The damping constant which corresponds to the emission of radiation from the state F is obtained by using the formula for Einstein's emission probability and is

$$\gamma_E = (C/R) \left| \int_0^\infty Ffr dr \right|^2, \tag{18}$$

where

$$C = \frac{32\pi^3 e'^2 \nu^3}{3hc^3} \frac{L + \frac{1}{2} \pm \frac{1}{2}}{2L + 1}, \tag{18'}$$

the upper sign applying to jumps $L \to L + 1$ and $e' \sim e/2$ is the effective charge of the neutron nucleus system. As $R \to \infty$, both $\Delta \nu$ and γ_E decrease towards zero but their ratio remains constant. The cross-section due to capture computed directly from the emission probability is

$$\sigma_C' = (2L + 1)\Lambda^2 \gamma_E/\Delta \nu. \tag{19}$$

If this expression approaches Λ^2 then $\gamma_E/\Delta\nu$ becomes comparable with unity and the levels are close enough together to make eq. (19) meaningless. It is then necessary to take into account the mutual influence of neighbouring levels. This can be done by means of the damping matrix.[14] The successive states of angular momentum $L\hbar$ will be denoted by indices j, l and their probability amplitudes by a_j. These satisfy

$$\left(\frac{d}{2\pi i dt} + \nu_j\right)a_j = i\Sigma \gamma^{jl} a_l, \qquad (20)$$

where

$$\gamma^{jl} = CJ_j^* J_l / R; \quad J_j = \int F_j f dr. \qquad (20')$$

In our case only states with the same magnetic quantum number can interact so that a complete specification of the states is obtained through their energy. Solutions of (20) in which all quantities vary as $\exp\{-2\pi i(\nu_0 - i\gamma)t\}$ correspond as closely as possible to the notion of a stationary state decaying under influence of radiation damping. From (20) one obtains

$$(\nu_j - \nu_0 + i\gamma)a_j = i\Sigma \gamma^{jl} a_l. \qquad (20'')$$

These equations with the complex eigenwert $\nu_0 - i\gamma$ can be reduced making use of the fact that γ^{jl} is a matrix of rank 1. Thus eliminating the a_l one finds

$$1 = \frac{iC}{R} \Sigma \frac{|J_j|^2}{\nu_j - \nu_0 + i\gamma} \qquad (21)$$

for the secular equation which determines ν_0 and γ. This equation will be solved approximately for the case of sharp resonance. The resonance will be supposed to take place at an energy $h\nu_F$ and to have a "half-value breadth" $2h\Gamma_F$. Close to resonance

$$|J_j|^2 = \frac{\Gamma_F^2 |I|^2}{(\nu_j - \nu_F)^2 + \Gamma_F^2}, \qquad (22)$$

where $|I|^2$ is the maximum value of $|J|^2$. This approximation will usually apply only in a region of a few Γ_F. The value of Γ_F can be estimated using[15]

$$4\pi\Gamma_F = v_r / \int \bar{G}^2 dr, \qquad (23)$$

where $\bar{G}$ is F for resonance, v_r is the velocity at resonance, and the integration is to be carried through the range of large values of $\bar{G}$. The quantity Γ_F is analogous to Γ_s of section (a). The state represented by $\bar{G}$ is analogous to the quasi-stationary state of section (a). In order to bring out the analogy we introduce a damping constant similar to the previous Γ_r

$$\Gamma_R = C|I|^2/2\int \bar{G}^2 dr = 2\pi C|I|^2 \Gamma_F/v_r, \qquad (23')$$

which is the damping constant of the state represented by $\bar{G}$ when that state is normalized within the nucleus and its immediate vicinity. Substituting (23') into (21), replacing the sum by an integral everywhere except in the vicinity of

ν_0 and performing the summation in that region on the assumption that the $\Delta \nu$ can be considered as equal to each other in that region gives

$$1 = i\{a \cot\left[\frac{\pi(\nu_{j0} - \nu_0 + i\gamma)}{\Delta\nu_{j0}}\right] + ia + \int_0^\infty \frac{\nu_r |J_j|^2 \Gamma_R d\nu_j}{\pi \nu_j |I|^2 \Gamma_F(\nu_j - \nu_0 + i\gamma)} \quad (24)$$

$$a = \left[\frac{\nu_r |J_j|^2 \Gamma_R}{\nu_j |I|^2 \Gamma_F}\right]_{\nu_0},$$

where the integral must be extended over all ν_j and the region around ν_0 is integrated over the real axis. The quantity ν_{j0} is any one of the ν_j located so close to ν_0 that the variation in $\Delta\nu$ in between can be neglected. In the approximation of eq. (22) the integration over the resonance region Γ_F leads to an equation which to within a sufficient approximation reduces to

$$1 + itTh = (Th + it)(a + ib); \quad b = \frac{\nu_{j0}|I|^2 q}{\nu_r |J_{j0}|^2(1 + q^2)} \quad (25)$$

with

$$\nu_0 - \nu_F = q\Gamma_F; \quad Th = \tanh\frac{\pi\gamma}{\Delta\nu_{j0}}; \quad t = \tan\frac{\pi(\nu_0 - \nu_{j0})}{\Delta\nu_{j0}}. \quad (25')$$

By eliminating t

$$Th + 1/Th = a + 1/a + b^2/a. \quad (25'')$$

which has the approximate solution

$$1/Th = a + 1/a + b^2/a, \quad (26)$$

For values of ν_0 which lie in the region where eq. (22) applies and where $\Delta\nu_{j0} \sim \Delta\nu_r$ we have approximately

$$\frac{\pi\gamma}{\Delta\nu} = \frac{\Gamma_R \Gamma_F}{\Gamma_R^2 + \Gamma_F^2(1 + q^2)}, \quad (26')$$

where it is supposed that $\Gamma_R \gg \Gamma_F$. If, however, $\Gamma_F \gg \Gamma_R$ then

$$\frac{\pi\gamma}{\Delta\nu} = \frac{\Gamma_R}{\Gamma_F(1 + q^2)}, \quad (26'')$$

which is equivalent to using the γ_E of eq. (18), (19); in this case one may compute using emission probabilities. If one is so far away from resonance that $b^2/a < a$, $1/a$ eq. (25'') gives

$$\gamma = \gamma^{j_0 j_0}, \quad (26''')$$

provided the right side is $\ll 1$. Here again the simple emission point of view applies. For $\Gamma_R \gg \Gamma_F$ all regions are approximated by

$$\frac{\pi\gamma}{\Delta\nu_{j0}} = \frac{\nu_r |J_{j0}|^2 \Gamma_R \Gamma_F(1 + q^2)}{\nu_{j0}|I|^2[\Gamma_R^2 + \Gamma_F^2(1 + q^2)]} \quad (27)$$

The treatment of scattering by means of the damping matrix is somewhat involved and will not be reproduced here. The phase shift due to $v_0 - v_{J0}$ when added to the phase shift already present in F_{J0} gives the phase shift required. The scattering is diminished by Γ_R in much the same way as it was diminished by it in section (a). By comparing (27) with (19)

$$\sigma_C = (2L+1) \frac{\Lambda^2 v_r |J_{J0}|^2 [\Gamma_R \Gamma_F (1+q^2)]}{\pi v_{J0} |I|^2 [\Gamma_R{}^2 + \Gamma_F{}^2 (1+q^2)]}, \tag{28}$$

which is similar to eq. (14), close to resonance. The factors $|J|^2/|I|^2$ and v_r/v_{J0} take into account the deviations from the dependence of $|J|^2$ on v given by (22). In (14) this is analogous to the dependence of Γ_s on v_{J0} combined with Dirac's frequency shift.

(c) Sharpness of resonance for single-body problem

The upper limit of integration in eq. (23) has been left indefinite. By Green's theorem

$$\frac{d}{dr}\left[F_1 \frac{dF_2}{dr} - F_2 \frac{dF_1}{dr}\right] + \frac{2\mu}{\hbar^2}(E_2 - E_1) F_1 F_2 = 0,$$

where F_1, F_2 correspond to energies E_1, E_2 and need not be regular at $r = 0$. Hence[16]

$$\frac{\partial}{\partial r}\left[F^2 \frac{\partial}{\partial E}\frac{\partial F}{F \partial r}\right] + \frac{2\mu}{\hbar^2} F^2 = 0. \tag{29}$$

In this section let F_i be the function inside the nucleus, and let F stand for the regular solution of the wave equation for $r \times$ radial function on the absence of the nuclear field. The normalization of F is such as to make it a sine wave $\sin(kr + \varphi)$ of unit amplitude at ∞. Similarly G is defined as satisfying the same differential equation as F but it is to be 90° out of phase with F at ∞, i.e. $\cos(kr + \varphi)$. The regular solution of the differential equation in the presence of the nuclear field, normalized in the same way as F and G, will be called $\bar{F}$. At the nuclear radius r_0

$$\bar{E}^2 = G^2 \bigg/ \left\{\left[G^2 \left(\frac{G'}{G} - \frac{F_i'}{F_i}\right)\right]^2 + \left[FG\left(\frac{F'}{F} - \frac{F_i'}{F_i}\right)\right]^2\right\}. \tag{30}$$

Here the accent stands for differentiation with respect to kr. At resonance $F_i'/F_i = G'/G$ and the second term in the curly bracket is then 1, while the first term is zero. As E changes to either side of the resonance value E_r the first term may become 1 for $E = E_r \pm \Delta E$ where ΔE is properly chosen. The half-value breadth is then $2\Delta E$ and $\Delta E = \hbar \Gamma_F$. The value of ΔE can be estimated by

$$\Delta E \frac{\partial}{\partial E}\left[G^2\left(\frac{G'}{G} - \frac{F_i'}{F_i}\right)\right]_{r_0} = 1. \tag{31}$$

Using eq. (29) and calculating the $\partial/\partial E$ for $E = E_r$ one obtains a result which can be expressed in terms of integrals up to R where R is any value of r which is greater than r_0. The function which is G for $r > r_0$ and $F_i(G/F_i)_{r_0}$ for $r < r_0$ is continuous at r_0 and at resonance its derivative with respect to r is also continuous. The function will be called $\bar{G}$ for $0 < r < \infty$. We have then[17]

$$\frac{E}{\Delta E} = k \int_0^R \bar{G}^2 dr + \left[\frac{G^2 E \partial}{k \partial E}\frac{\partial G}{G \partial r}\right]_{r=k}; \quad \Delta E = h\Gamma_F. \tag{32}$$

The right side of this result is independent of R and is finite. The term outside the integral should be included in eq. (23) changing

$$\int G^2 dr \to \int_0^R \bar{G}^2 dr + \left[\frac{G^2 E \partial}{k^2 \partial E}\frac{\partial G}{G \partial r}\right]_{r=k}. \tag{32'}$$

Equation (32) has a well-defined meaning only if resonance is sharp. Otherwise the $\partial/\partial E$ entering in eq. (31) cannot be supposed to be sufficiently constant through the half-breadth $2h\Gamma_F$. It cannot be expected to hold for the broad S resonance discussed by Bethe.

(d) CAPTURE BY p STATES

For a potential well of constant depth

$$F_i = \sin z/z - \cos z; \quad F = \sin \rho/\rho - \cos \rho;$$
$$G = \cos \rho/\rho + \sin \rho, \tag{33}$$

where

$$z = Kr; \quad \rho = kr; \quad K = \mu v_i/\hbar; \quad k = \mu v/\hbar \tag{33'}$$

v_i, v being, respectively, the velocities inside and outside the nucleus. The resonance condition is

$$z \sin z/[\sin z/z - \cos z] = \rho \cos \rho/[\cos \rho/\rho + \sin \rho].$$

For slow neutrons $\rho \ll 1$ the right side is $\ll \rho^2$ and therefore very small. The first resonance point is obtained for $z = \pi - \epsilon$, $\epsilon \sim \rho^2/\pi$. It will suffice to take $z = \pi$. By substituting into eq. (32) it follows that

$$\Delta E/E = 2\rho/3 = 4\pi r_0/3\Lambda. \tag{33''}$$

For $E = (1/40)$ volt, $\Lambda = 1.8 \times 10^{-8}$ cm, $h\Gamma_F = 5.8 \times 10^{-6}$ volt.

For 3-MeV γ-rays a reasonable value of $h\Gamma_R$ is 5.8 volts. The cross-section at resonance is by eq. (28) $3\varLambda^2\Gamma_F/\pi\Gamma_R = 300 \times 10^{-24}$ cm². Since scattering is of the order Γ_F/Γ_R times capture the scattering cross-section is small. According to eq. (33″) the cross-section at resonance for p terms with $\Gamma_R \gg \Gamma_F$ can be expected to vary as v and $h\Gamma_F$ as v^3. The range in which p terms can be expected to give large capture cross-sections and small scattering is therefore roughly from 1/40 volt to 1 volt. At higher velocities $h\Gamma_F$ is likely to be higher than $h\Gamma_R$. In the absence of an apparent reason for nuclear p levels to fall in this narrow velocity range, an explanation in terms of p terms although possible is improbable on account of the small range of neutron velocities required.

3. Capture Through s Wave

(a) This section will contain the calculation of the A_s used in 2(a). It is supposed that the system "nucleus+neutron" can be treated in first approximation by means of an effective central field acting on the neutron. The difference between the Hamiltonian of the system and the Hamiltonian corresponding to the central field will be called H'. On account of this difference there exist transitions from the s wave of the incident state to quasi-stationary excited states of the "nucleus+neutron" system. Normalizing the s waves within a sphere of radius R the wave function inside the nucleus is

$$C \sin Kr/r; \quad C^{-2} = [1 + (U/E) \cos^2 Kr_0]2\pi R,$$

where
$$K^2/k^2 = (U + E)/E$$

and U is the depth of the potential hole. The interaction energy H' involves besides r also internal coordinates x. The wave function of the whole system in the incident state may be written $C\psi_0(x) \sin Kr/r$ and in the quasi-stationary state $\psi_Q(r, x)$. The matrix element M_s of eq. (1) is then

$$M_s = \int \psi_Q(r, x) H' C \sin Kr \psi_0(x) dv/r, \tag{34}$$

where dv is the volume element of the whole system. The state Q is by definition such that the integral of $|\psi_Q|^2$ through nuclear dimensions is unity. The order of magnitude of M_s is therefore

$$M_s = C\bar{H}r_0^{\frac{1}{2}}, \qquad (34')$$

where $\bar{H}$ is an average of H' through the nucleus and may have reasonably a value of 0.5 MeV. It cannot be specified further without detailed calculation which would probably be unsatisfactory in the present state of nuclear theory. Since $\Delta E = hv/2R$,

$$h\Gamma_s \cong \frac{\bar{H}^2 r_0}{2 A U \cos^2 K r_0}. \qquad (34'')$$

According to eq. (14)

$$\sigma_c = \frac{A r_0}{2\pi} \frac{\bar{H}}{U \cos^2 K r_0} \frac{\bar{H} h \Gamma_r}{h^2(\Gamma_r + \Gamma_s)^2 + (E - E_r)^2}, \qquad (35)$$

where E_r is the value of E for resonance. According to this formula there are two maxima for σ_c, one for $E = E_r$ and one for $E = 0$. The expected cross-sections are given in Table 1 to about 10 per cent accuracy. The numbers correspond to $A(kT) = 1.8 \times 10^{-8}$ cm; $A(1\text{ volt}) = 2.9 \times 10^{-9}$ cm; $r_0 = 3 \times 10^{-13}$ cm; $U\cos^2 Kr_0 = 10^7$ volts. For $E_r = 1/40$, 1 volt the table shows large cross-sections at thermal energies and above. The condition is similar to Bethe's except for a relatively sharper resonance determined by $h\Gamma_r$. For 50 volts one sees the development of two maxima one at resonance and one at thermal energies. For $E_r = 1000$ and 10,000 volts the maximum at thermal energies decreases as E_r^{-2} and the maximum at resonance roughly as $E_r^{-\frac{1}{2}}$. For such high values of E_r scattering has a chance of becoming comparable with absorption or even greater than the absorption at resonance. In the thermal energy region the $1/v$ law is obeyed for high values of E_r; for low E_r the maximum at E_r interferes with the $1/v$ law and the region of its validity is displaced below thermal energies.

In Table 1 only the effect of a quasi-stationary level at E_r is considered. In addition there may be effects of other levels as well as

CAPTURE OF SLOW NEUTRONS

TABLE 1. CALCULATED CROSS-SECTIONS FOR NEUTRON CAPTURE

Position of resonance (volts)	$\bar{H}$ (MeV)	$h\Gamma_r$ (volts)	$h\Gamma_s$ at resonance (volts)	$10^{24}\sigma$ (cm²) Resonance	$10^{24}\sigma$ (cm²) Thermal energies
1/40	0.1	10	0.01	90,000	
	0.1	1	0.01	900,000	
1	0.1	10	0.05	14,000	
	0.1	1	0.05	140,000	
50	0.1	10	0.37	2000	3500
	0.5	10	9	13,400	80,000
	0.1	1	0.37	11,000	350
	0.5	1	9	4800	9000
1000	0.1	10	1.6	320	9
	0.5	10	40	420	200
10000	0.1	10	5	60	0.09
	0.5	10	125	18	2

radiation jumps of the kind considered by Bethe and Fermi which do not depend on the existence of virtual levels. It is thus probable that in most cases there is a region with a $1/v$ dependence although it may be at times masked by a resonance region.

(b) DIRAC'S FREQUENCY SHIFT

In the above estimates the effect of Dirac's frequency shift was neglected. This is given by

$$(h\Delta\nu)_D = \int_0^\infty \frac{h\Gamma_s}{\pi} \frac{dE}{E - E_0} = \frac{\bar{H}^2 r_0}{\pi \Lambda_0 x_0} \int_0^\infty \frac{x^2 dx}{(x^2 - x_0^2)(x^2 + a^2)},$$

where $x = E^{\frac{1}{2}}$, $a^2 = U \cos^2 Kr_0$ and the value of $h\Gamma_s$ was substituted by means of eq. (34″). Here the subscript 0 refers to the neutron energy E_0 and the principal value of the integral is understood. Evaluating the expression

$$(h\Delta\nu)_D = \frac{\bar{H}^2 r_0 U^{\frac{1}{2}} \cos Kr_0}{2\Lambda_0 (E_0 + U \cos^2 Kr_0) E_0^{\frac{1}{2}}}. \tag{36}$$

The shift is seen to be of the order of 3000 times $h\Gamma_s$ for $E_0 = 1$

volt. The shift is nearly independent of the velocity. In the approximation of eq. (36)

$$\frac{d(h\Delta\nu)_D}{dE_0} = \frac{h\Gamma_s}{E_0} \frac{E_0^{\pm}}{U^{\pm} \mid \cos Kr_0 \mid}, \tag{36'}$$

which shows that the variation in the shift is small and of the order of $2 \times 10^{-5}(E - E_r)$ for $\bar{H} = 0.1$ MeV.

4. Discussion

(a) Absence of scattering

According to Dunning, Pegram, Fink and Mitchell[18] the elastic scattering of slow neutrons by Cd is less than one per cent of the number captured. According to A. C. G. Mitchell and E. J. Murphy[19] scattering as detected by silver is about the same as absorption for Fe, Pb, Cu, Zn, Sn while for Hg scattering is about 1/80 of the absorption. In the later communication of Mitchell and Murphy[19] it is also found that Ag, Hg, Dd are poor scatterers of slow neutrons detected by silver. It is interesting that Ag shows small scattering in these experiments because the detection took place by means of silver and that Hg and Cd show small scattering because they have large absorption cross-sections.[18] The observation of scattering by a material having large absorption is difficult because the neutrons entering the material are absorbed before they can be scattered and it is possible that to some extent the failure to observe scattering in good absorbers is due to this cause. The absence of observed scattering in the region of strong absorption is therefore not a surprise, particularly in view of the relatively small number of neutrons available for experimentation. It seems more significant, however, that strong absorbers do not show, so far, strong scattering in any velocity region because, according to the Fermi–Bethe theory, the scattering cross-section should be large in a wide range of energies. The experimental evidence says little about the ratio of scattering to absorption near resonance. It indicates that this ratio is less than 1/10 in most cases. It is impossible, therefore, to ascertain definitely the ratio Γ_s/Γ_r until more detailed experimental data are available. According to Table 1 the condition $\Gamma_s/\Gamma_r < 1/10$ can be satisfied in many ways up to velocities of over 1000 eV.

(b) Magnitude of interaction with internal states and probability of internal state in required region

In Table 1 arbitrary assignments of values of Γ_s, Γ_r were made. It will be noted that at low neutron velocities the desired large capture cross-sections are easily obtained through relatively wide bands having a half-value breadth $2\Gamma_r$. Keeping Γ_r fixed one can decrease the interaction energy $\bar{H}$ to 10,000 eV for $h\Gamma_r = 1$ volt, $E_r = 1$ volt and still have a cross-section of 1000×10^{-24} cm² in an energy range up to 2 volts. In some cases relatively weak radiative transitions will come into consideration leading to smaller Γ_r. For such transitions $\bar{H}$ need not be as large as 10,000 eV in order to have cross-sections of 1000×10^{-24} cm² in the resonance region. For the large energies involved in nuclear structure it is reasonable to expect interaction energies of the order of 10,000 volts between practically any pair of levels not isolated by a selection rule and interaction energies of the order 100,000 volts between a great many levels.

There are about ten elements among 72 observed that show cross-sections of more than 500×10^{-24} cm². Allowing for the fact that there are more isotopes than elements it appears fair to say that the chance of such an anomalous cross-section is about 1/20. One can try to account for these solely by the low-velocity regions which exist for any resonance level, thus probably overestimating the necessary number of levels. In order that $\sigma_c > 500 \times 10^{-24}$ cm² at 1/40 volt for a nucleus having $r_0 = 10^{-12}$ cm and $h\Gamma_r = 10$ volts the resonance region must be not farther than at $|E_r| = \bar{H}/420$ from thermal energies by eq. (35). We do not wish $h\Gamma_s$ at thermal energies to be greater than 0.1 volt so as not to have too much scattering and therefore $\bar{H}$ should be below 2×10^5 eV at the higher E_r. Thus E_r should be kept below about 460 volts in order to give the large capture cross-sections for $E = 1/40$ volt together with small scattering. A level below ionization will also be effective in producing an increased absorption. The observed number of large absorptions corresponds in this way to one level in 900 volts for 1/20 of nuclei or one level every 18,000 volts for a single nucleus. In addition some cross-sections will be caused by direct resonance. Just how many is uncertain but it is clear that such effects exist in Cd, Ag, Au, Rh, In.

The average spacing between the γ-ray levels of Th C" as given in Gamow's book is about 100,000 volts and this is apparently the order of magnitude usual for γ-ray levels of radioactive nuclei. There appears to be no reason why the energy levels found through the analysis of γ-ray spectra should include all the nuclear levels and there may be as many as one level in 20,000 of a kind that may be responsible for coupling to incident neutrons. It should be remembered here that some of the levels may be active even though the coupling is weak so that more possibilities are likely to matter than for the γ-rays of radioactive nuclei.

For a complicated configuration of particles it seems reasonable to consider a total number of 100 possible levels per configuration because protons and neutrons can be combined separately to give different states. On this basis we deal with an average spacing between configurations of about 2 MeV which is not excessively small. It is, of course, impossible to prove anything definitely without calculating the levels; this appears to be premature at present on account of uncertainties in nuclear theories.

(c) EXISTENCE OF TWO MAXIMA

According to the calculation given above it is expected that there will be in general two maxima one of which should be at resonance and another at $v = 0$. According to the experiments of Rasetti, Segrè, Fink, Dunning and Pegram[20] the $1/v$ law is not obeyed by Cd but is obeyed by Ag. Cadmium has therefore a resonance region close to thermal velocities. In the classification of Fermi and Amaldi[21] this region must be affected by the C group since absorption measurements by the Li ionization chamber which was used in these experiments agree for most elements with the measurements of Fermi and Amaldi on the C group.[22] The verification of the $1/v$ law for Ag by the rotating wheel indicates that in Ag the resonance band is located above thermal energies. This conclusion is in agreement with the smallness of the temperature effect for the A neutrons detected by silver which was recently established by Rasetti and Fink.[23] Since Rh behaves similarly to Ag in these temperature experiments Rh also has a resonance region above thermal velocities. Fermi and Amaldi have evidence that D neutrons, which affect Rh, are different from A neutrons which affect Ag. It is very probable that both of these groups lie above the thermal region and they may reasonably cover a range of 30 volts inasmuch as the B group overlaps weakly with both A and D.

According to Szilard[6] In shows strong selective effects outside the C group and according to Fermi and Amaldi[21] the same period of In (54 min) detects the D group. The number of neutrons in the groups is presumably in the ratios $C/80 = B/20 = D/15 = A/1$. One could try to conclude that the order of increasing energies is C, B, D, A on the assumption that the number of neutrons increases towards low energies. Such a conclusion is dangerous because little is known about the velocity distribution, because within each group there may be several bands at different velocities, and also because the number of expected neutrons in a group should depend on its width. Temperature effects show that practically all captures increase as the energy is lowered. The effects are strongest[24] for Cu, V, are smaller for Ag, Dy, weaker for Rh and weakest for I. The absorption coefficient for C neutrons is, however, larger for Rh than for Ag indicating that the smaller temperature effect in Rh is due to a relatively greater importance in it of a band above thermal energies. All temperature effects agree in indicating the presence of a region in which the $1/v$ law is followed approximately but again no definite conclusion about the order of bands is possible. The low temperature effect in I would tend to indicate that its absorption region is high and detection-absorption experiments on I and Br tend to indicate that their bands are isolated from the others discussed here; perhaps these isotopes have resonance bands at higher energies. A new band was recently discovered in Au by Frisch, Hevesy and McKay[25] which represents strong absorption on a weaker background. The large number of selective effects observed makes the present explanation reasonable and the existence of a region of low energies in which the absorption decreases with energy is seen to fit well with expectation.

(d) OTHER POSSIBILITIES

One may consider weak long range forces as a possible explanation of the same phenomenon. Potentials of the order of neutron energies in a region comparable with the neutron wave-length would produce strong effects on absorption and scattering. For thermal energies the wave-length is of atomic dimensions and one would therefore expect the binding energy of deuterium compounds to be different from that of hydrogen compounds by an amount comparable to 1/40 volt if such potentials were present. Such energy differences do not exist. It would be possible to devise potentials which fall off sufficiently rapidly with distance to make the interaction potential negligible for chemical binding and which would cover a total region appreciably larger than the nucleus. Such hypotheses seem improbable without additional argument. Besides, special relations between the phase integrals through the nuclear interior and the part of the range of force outside the nucleus would have to be set up in order to make absorption large and scattering small. It is improbable that the large number of bands could be accounted for by any single particle picture.

Forces between electrons and neutrons even though they may exist are not likely to have much to do with the bands. Thus it has been shown by Condon[26] that electron–neutron interactions would give rise to scattering cross-sections varying roughly as the square of the atomic number Z on the assumption that the electron–neutron forces alone are responsible for the scattering. Forces inside the nucleus must also be supposed to contribute to the phase shifts responsible for scattering. Since these forces also vary with Z one could obtain a more complicated dependence of the scattering cross-section by suitably adjusting the nucleus–neutron and electron–neutron potentials. On such a picture one could try to account for sharp resonances by making the electron–neutron interaction repulsive. However, Condon's calculation shows that isotope shifts would be also produced by these interactions. It is improbable that the isotope shift is due solely to neutron–electron interaction because the deviation from the inverse square law inside the nucleus due to smearing out of protons produces a considerably larger effect than the observed shift. But it would also be unreasonable to try to combine the proton and neutron effects in the nucleus so as to have each large but their difference small. It is therefore probable that the electron–neutron interaction is not much larger than that which corresponds to the observed isotope shift. Since the density of the Fermi–Thomas distribution varies for small r as $r^{-\frac{3}{2}}$ the effective potential acting on the neutron will become high for small r. However, calculation shows that it is not high in a wide enough region to account for sharp resonances if the limitation due to the isotope shift is considered.

Bombardment of light nuclei with charged particles has also shown the existence of resonances. Thus there are resonances[27] for the emission of γ-rays in proton bombardment of Li, C, F, and similarly there are the well-known resonances in disintegrations produced by α particles. Experimental methods have not been very suitable so far for the detection of resonance regions on account of the scarcity of monochromatic sources and the necessity of using thin films. In Li protons are apparently able to produce γ-rays in two ways; by resonance at 450 kV and by another process at higher energies. In fluorine there are several peaks. In carbon there was an indication of the main resonance peak being double. It appears possible that many more levels will be detected inasmuch as neutron experiments indicate a high density of levels. Calculations on the radiative capture of carbon under proton bombardment[15] lead to a higher yield than is observed by a factor of several thousand. In these calculations the capture was supposed to occur by a jump from the p state of the incident wave to an s state of the N^{13} nucleus. The calculated half-value breadth due to proton escape from the quasi-stationary p level was of the order $h\Gamma_F \sim 10{,}000$ eV and thus much larger than the width due to radiation damping. The yield in thick targets under these conditions is nearly independent of the special value of $h\Gamma_F$. It is clear from the formulas given here for neutron capture that one can decrease the theoretically expected yield either by ascribing the capture to a transition having a small probability of radiation (small Γ_r) such as would correspond to quadrupole or other forbidden transitions or else by using an intermediate state of excitation of the nucleus with a small transition probability to the incident state of the proton (small Γ_s). In the latter case this transition probability would have to be made so small as to have $\Gamma_s < \Gamma_r$ and the observed width of resonance would have to be ascribed to experimental effects. If $\Gamma_s < \Gamma_r$ the thick target yield depends on Γ_s and is proportional to it for small Γ_s. The apparent disagreement between theory and experiment previously found for carbon is thus not alarming from the many-body point of view and supports the belief that excitation states of the nucleus have often to do with the simultaneous excitation of more than one particle.

The excitation states responsible for the neutron absorption bands make it possible for a fast neutron to lose energy by inelastic impact with the nucleus. Estimates show that the cross-sections for such processes are likely to be small when energy losses are high. The cross-section is estimated to be

$$\frac{\Lambda_1}{4\pi\Lambda_2} \frac{\overline{H}^2 r_0^2}{U^2 \cos^4 K r_0}, \tag{37}$$

where Λ_1, Λ_2 are, respectively, neutron wavelengths in the incident and final states. For large energy losses $\Lambda_2 \gg \Lambda_1$ and only a small effect need be expected. The excitation levels responsible for neutron capture will give small values Λ_1/Λ_2. Excitation levels located lower are more favourable and probably the excitation of Pb to about 1.5 MeV has to do with such a possibility.[28]

We are very grateful to Professors R. Ladenburg and F. Rasetti for interesting discussions of the experimental material.

Notes

[1] H. A. Bethe, *Phys. Rev.* **47**, 747 (1935). We refer to this paper as H.B. in the text.

[2] E. Amaldi, O. d'Agostino, E. Fermi, B. Pontecorvo, F. Rasetti and E. Segrè, *Proc. Roy. Soc.* **A149**, 522 (1935).

[3] Perrin and Elsasser, *Comptes Rendus* **200**, 450 (1935).

[4] Beck and Horsley, *Phys. Rev.* **47**, 510 (1935).

[5] Moon and Tillman, *Nature* **135**, 904 (1935); Bjerge and Westcott, *Proc. Roy. Soc.* **A150**, 709 (1935); Arsimovitch, Kourtschatow, Miccovskii and Palibin, *Comptes Rendus* **200**, 2159 (1935); Ridenour and Yost, *Phys. Rev.* **48**, 383 (1935); Pontecorvo, *Ricerca scientifica* **6-7**, 145 (1935).

[6] L. Szilard, *Nature* **136**, 950 (1935).

[7] J. H. Van Vleck, *Phys. Rev.* **48**, 367 (1935).

[8] O. K. Rice, *Phys. Rev.* **33**, 748 (1929); **35**, 1551 (1930); **38**, 1943 (1931); *J. Chem. Phys.* **1**, 375 (1933). A similar process was used by M. Polanyi and E. Wigner, *Zeits. f. Physik* **33**, 429 (1925).

[9] P. A. M. Dirac, *Zeits. f. Physik* **44**, 594 (1927).

[10] V. Weisskopf and E. Wigner, *Zeits. f. Physik* **63**, 54 (1930).

[11] V. Weisskopf, *Ann. d. Physik* **9**, 23 (1931).

[12] G. Breit, *Rev. Mod. Phys.* **5**, 91, 104, 117 (1933).

[13] Appendix I.

[14] G. Breit, *Rev. Mod. Phys.* **5**, 117 (1933); G. Breit and I. S. Lowen, *Phys. Rev.* **46**, 590 (1934).

[15] G. Breit and F. L. Yost, *Phys. Rev.* **48**, 203 (1935). See also eq. (32').

[16] J. A. Wheeler. We are indebted to Dr. Wheeler for communicating to us other applications of this relation.

[17] Cf. eq. (22) footnote 15. In calculations with Coulombian fields it is sometimes convenient to transform eq. (32) of the text into

$$\frac{E}{\Delta E} = G^2 \left[\frac{k}{F_j^2} \int_0^{r_0} F_i^2 dr - \frac{k}{F^2} \int^{r_0} F^2 dr - \frac{EG^2 \partial}{kr \partial E} \left(\frac{kr}{FG} \right) \right]$$

all quantities outside the integrals being taken for $r = r_0$.

[18] J. R. Dunning, G. B. Pegram, G. A. Fink and D. P. Mitchell. *Phys. Rev.* **48**, 265 (1935).

[19] A. C. G. Mitchell and E. J. Murphy, *Phys. Rev.* **48**, 653 (1935). Cf. also *Bull. Am. Phys. Soc.* **11**, paper 27, Feb. 4, 1936.

[20] F. Rasetti, E. Segrè, G. Fink, J. R. Dunning and G. B. Pegram, *Phys. Rev.* **49**, 103 (1936).

[21] E. Amaldi and E. Fermi, *Ricerca scientifica* **2**, 9 (1936); E. Fermi and E. Amaldi, *Recerca scientifica* **2**, 1 (1936).

[22] Unpublished results of F. Rasetti. We are very grateful to Professor Rasetti for informing us of these results.

[23] F. Rasetti and George A. Fink, *Bull. Am. Phys. Soc.* **11**, Paper 28, Feb. 4, 1936.

[24] P. B. Moon and R. R. Tillman, *Proc. Roy. Soc.* **A153**, 476 (1936).

[25] O. R. Frisch, G. Hevesy and H. A. C. McKay, *Nature* **137**, 149 (1936).

[26] E. U. Condon, in press. We are indebted to Professor Condon for showing us his manuscript before publication.
[27] L. R. Hafstad and M. A. Tuve, *Phys. Rev.* **48**, 306 (1935); P. Savel, *Comptes Rendus* **198**, 1404 (1934), Ann. de physique **4**, 88 (1935).
[28] J. Chadwick and M. Goldhaber, *Proc. Roy. Soc.* **A151**, 479 (1935).

Appendix 1

VARIATION OF DAMPING CONSTANT WITH ENERGY AND DIRAC'S FREQUENCY SHIFT

Equation (6) of the text lead to [cf. eqs. (126') to 129') of reference in note 12]

$$(\nu_{s0} - \nu_0 + i\gamma)(\nu - \nu_0 + i\gamma) = |A_{s0}|^2 \\ + (\nu_{s0} - \nu_0 + i\gamma)\left[\Sigma'|A_s|^2 \frac{1 - e^{2\pi i(\nu_0 - \nu_s - i\gamma)t}}{\nu_s - \nu_0 + i\gamma} \\ + \Sigma|B_r|^2 \frac{1 - e^{2\pi i(\nu_0 - \nu_r - i\gamma)t}}{\nu_r - \nu_0 + i\gamma}\right] \quad (38)$$

which determines Γ by comparison with (7) $[\Gamma \gg \gamma]$.

It is by no means natural that this equation can be satisfied because the right side depends on t. If the A_s as well as the B_r were all essentially equal and if γ were great in comparison with the frequency differences of consecutive levels the sums could be transformed into integrals in the well-known way[10-12] so that (9) as well as (6) would follow. We shall attempt here a more exact procedure.

Consider the Σ_s' in the square brackets. It is natural to divide the range of ν_s into two parts: one for which $|\nu_s - \nu_0| > a \gg \gamma$ and one for which $|\nu_s - \nu_0| \leq a$. Since $\overline{|A_s|^2}/\Delta\nu_s$ changes slowly this quantity will be replaced by a constant in $|\nu_s - \nu_0| \leq a$ and its value may be taken to be that at ν_0 for the evaluation of the contribution of this region. We have then to consider

$$(\overline{|A_s|^2})_{\nu_0} \sum_{\nu_0-a}^{\nu_0+a}{}' \frac{1 - e^{2\pi i(\nu_0 - \nu_s - i\gamma)t}}{\nu_s - \nu_0 + i\gamma}. \quad (39a)$$

An exact evaluation of this sum is not simple because γ and $\Delta\nu$ are of the same order of magnitude and the replacement of (39a) by an integral is somewhat objectionable. This point has never been completely cleared up and we have only qualitative arguments in favor of the correctness of the replacement of (39a) by an integral. For $\nu_s - \nu_0$ of the order of a few γ such a replacement is indeed meaningless but fortunately this region is not vital for $t \ll 1/\gamma$ since the numerator of (39a) is then small. For larger $|\nu_s - \nu_0|$ the terms of (39a) vary more smoothly and finally they become rapidly oscillating for $|\nu_s - \nu_0|t \gg 1$ which can be satisfied simultaneously with $t \ll 1/\gamma$ provided $a \gg \gamma$. The smallness of γ is thus not as serious as might appear from the fact that $\gamma/\Delta\nu \sim 1$. It should also be observed that the treatment of Rice[8] using real eigenwerte for a single one-dimensional continuum is in agreement with replacing (39a) by an integral. The result of doing so is given by (9).

In addition one has the contribution of $|\nu_s - \nu_0| > a$. This integral can be treated neglecting γ because it is of interest to evaluate γ only to quantities of order $\gamma/(\nu - \nu_0)$ and because the discussion is supposed to apply only to $\gamma t \ll 1$. This integration gives

$$\left(\int_0^{\nu_0-a} + \int_{\nu_0+a}^{\infty}\right) \frac{\overline{|A_s|^2}}{\Delta\nu_s} \frac{d\nu_s}{\nu_s - \nu_0} \tag{39b}$$

which means that the principal value of the $\int$ is understood. Similarly one obtains a contribution due to $|B_r|^2$. These two integrals give the Dirac frequency shift which is included in eq. (15).

As stated in the text the difference between Γ' and Γ does not affect the absorption for $t \gg 1/\Gamma$. Thus for the initial condition given by eq. (4)

$$|b_r|^2 = \frac{|B_r|^2 |A_{s0}|^2}{(\nu - \nu_0)^2 + \Gamma^2} \left\{ \frac{1 + e^{-4\pi\gamma t} - 2e^{-2\pi\gamma t} \cos 2\pi(\nu_r - \nu_0)t}{(\nu_r - \nu_0)^2 + \gamma^2} \right.$$
$$+ \frac{1 + e^{-4\pi\Gamma' t} - 2e^{-2\pi\Gamma' t} \cos 2\pi(\nu_r - \nu_0)t}{(\nu_r - \nu)^2 + \Gamma'^2} + \text{cross product term} \right\}.$$

Only the first fraction in the curly brackets contributes to the steady increase of $\Sigma |b_r|^2$ in times $\gg 1/\Gamma$. Its contribution is

$$\frac{\pi |A_{s0}|^2 \overline{|B_r|^2}}{\gamma[(\nu - \nu_0)^2 + \Gamma^2]\Delta\nu_r} (1 - e^{-4\pi\gamma t}).$$

The last factor is for practical purposes $4\pi\gamma t$. The second and third terms in the curly bracket give terms $\exp(-4\pi\Gamma' t)$, $\exp(-2\pi\Gamma' t)$ and constants. The first two kinds die off and the last kind represents the effect of transients which do not matter in the long run, so that for times not too large as compared with $1/\gamma$ and yet great as compared with $1/\Gamma$ one may consider the rates of change of $\Sigma |b_r|^2$ and of $\Sigma' |a_s|^2$ to be $4\pi\gamma\Gamma_r/\Gamma$ and $4\pi\gamma\Gamma_s/\Gamma$. These are the results used in the text.

5

Fluctuations of Nuclear Reaction Widths[†]

C. E. PORTER AND R. G. THOMAS

Abstract. The fluctuations of the neutron reduced widths from the resonance region of intermediate and heavy nuclei have been analyzed by a statistical procedure which is based on the method of maximum likelihood. It is found that a chi-squared distribution with one degree of freedom is quite consistent with the data while a chi-squared distribution with two degrees of freedom (an exponential distribution) is not. The former distribution corresponds to a Gaussian distribution for the reduced-width amplitude, and a plausibility argument is given for it which is based on the consideration of the matrix elements for neutron emission from the compound nucleus and of the central limit theorem of statistics. This argument also suggests that within the framework of the compound-nucleus theory all reduced-width amplitudes have Gaussian distributions, and that many of the distributions for the various channels may be independent. One consequence of the latter suggestion is that the total radiation width for a given spin state which is formed in neutron capture will be essentially constant, in agreement with some observations, because it is the sum of many partial radiation widths. The fluctuations of the provisional fission widths of U^{235} are best described by a chi-squared distribution with about $2\frac{1}{3}$ degrees of freedom, indicating that there are effectively only a few independently contributing fission channels.

I. General Remarks

Several hundred resonances have been observed in the Brookhaven fast chopper work on total neutron cross-sections of intermediate and heavy nuclei in the neutron energy range up to several hundred electron volts.[1] For many of these resonances it has been possible to deduce the neutron width Γ_n and the velocity-indepen-

[†] *Phys. Rev.* **104**, 483 (1956). Work performed under the auspices of the U.S. Atomic Energy Commission.

dent reduced width $\Gamma_n{}^0 = \Gamma_n/E_0{}^{\frac{1}{2}}$, where E_0 is the resonance energy.[2,3] In a typical sample of from ten to fifteen resonances the reduced widths are observed to fluctuate violently, the ratio of the largest to the smallest being as high as several hundred. Indeed, Hughes and Harvey[4] have recently shown that the aggregate of the reduced-width data for fourteen nuclides is reasonably consistent with exponential-like distributions, one of the form $x^{-\frac{1}{2}} \exp(-\frac{1}{2}x)$ and another of the form $\exp(-x)$, where $x = \Gamma_n{}^0/\langle\Gamma_n{}^0\rangle_{\text{Av}}$. In view of the importance to nuclear reaction theory and to nuclear engineering of knowing which of the two distributions is more likely to be the correct one, we have made a more quantitative statistical analysis of the data. This analysis shows that the former distribution is quite consistent with the data, whereas to the latter one it assigns a very small probability of being correct. The most significant consideration of our analysis which enables this distinction to be made is the accounting for the possibility that levels with small widths, of which there are predicted to be a relatively large number in the former distribution, will not be observed. A second consideration which also enhances this distinction is the accounting for the errors introduced when finite-sample averages are used as estimates for the infinite-sample (population) averages $\langle\Gamma_n{}^0\rangle_{\text{Av}}$.[5]

The $x^{-\frac{1}{2}} \exp(-\frac{1}{2}x)$ distribution is also more reasonable on theoretical grounds. It corresponds to a Gaussian distribution for the reduced-width amplitude (that is, for $x^{\frac{1}{2}}$) for neutron emission from the compound nucleus while the $\exp(-x)$ reduced-width distribution corresponds to an amplitude distribution which assigns a zero probability for a zero amplitude. The reduced-width amplitude is proportional to an integral (matrix element) of the product of a wave function for the state of the compound nucleus and a wave function for the neutron channel, the integral being over the nuclear configuration space of many dimensions (essentially $3A$, where A is the number of nucleons).[6] In the spirit of the compound-nucleus theory, the former wave function is presumed to be very complex as a result of the strong nuclear interactions, and the wave functions for the various states are presumed to be essentially unrelated to each other. One may regard the matrix element as composed of contributions from many "cells" of the configuration space, the sign of the contribution from a particular cell being positive with the same probability that it is negative, and the sign and magnitude of a particular contribution being random from level to level and independent of the signs and magnitudes of the contributions from the other cells. The over-all size of each cell may be supposed to be such that each linear dimension is about 1×10^{-13} cm, the characteristic wavelength of a nucleon in the nucleus, so that in a heavy nucleus there will be a very large number of independently-contributing cells. In consideration of the central-limit theorem of statistics,[7] it may be expected that the probability distribution for the matrix elements (that is, for the sum over cells) will be approximately normal (that is, Gaussian) with zero mean and asymptotically normal in the limit as the number of effective independently-contributing cells becomes infinitely large (as in a hypothetical, infinitely heavy nucleus).[8]

The above arguments are not intended to constitute a derivation for the normal distribution but they do make it a plausible one. Departures from normality are to be expected. For example, according to the Wigner limit the reduced widths $\Gamma_n{}^0$ cannot exceed several thousand electron volts. However, this limit is millions of times larger than typical average reduced widths of heavy nuclei, so that the truncation occurs far out in the tail of the distribution and will not significantly affect the present considerations.[9]

It seems reasonable to hypothesize that all reduced-width amplitudes for levels of the compound nucleus (that is, levels of fairly high excitation) will be distributed approximately normally with zero means when sampled from level to level;[10] the variances of these distributions are just the average reduced widths. The amplitudes associated with the *partial* radiative capture widths for the compound states will also be assumed to have such a distribution because these amplitudes are proportional to matrix elements involving the wave functions of the compound states (as well as to wave functions for the final states). On the other hand, the distribution for the *total* radiation width for a particular spin state which

FLUCTUATIONS OF NUCLEAR REACTION WIDTHS 195

is involved in neutron capture is expected to be very narrow, its variance being inversely proportional to the number of contributing partial widths. However, this demonstration (section III) involves the additional assumption that the distributions for the various partial widths are independent,[11] which assumption is more open to question than is the normality assumption. Although the independence assumption may be applicable to the bulk of the transitions to states of fairly high excitation, which states are dense and presumably complex like the states of the compound nucleus, it may not be applicable to the direct transitions to low-lying states which may not be sufficiently complex. That is, several low-lying states might be identical in all respects except for a single property which is not involved in the radiative matrix elements except possibly through a common factor; the distributions of the matrix elements to these several states would then be completely correlated, although they could still be normal. However, for the treatment of III such transitions are combined as a single "independently-contributing" transition. The above-mentioned variance of the distribution for the total radiation width is then inversely proportional to the number of "independently-contributing" partial widths rather than to the actual number of partial widths. It is hard to say just what difference is to be expected of these two numbers; it would depend on the relative importance of the contributions from transitions to the low-lying states compared to the contributions from the states of fairly high excitation.

The distribution of fission widths is considered in section IV. Offhand, one might expect this distribution to be very narrow because, like radiative capture, there are many final states for the fission process, one for each possible fragment pair and additional ones for each possible pair of excitation states that are energetically accessible. The amplitudes associated with the partial widths for these states are expected to have normal distributions. However, if in the fission act the nucleus passes through only one or a few well-ordered nuclear states (fission channels) which describe the saddle-point configuration,[12] the various partial widths would be highly correlated (see the appendix). Indeed, if there were only one such state, the partial width distributions would be completely correlated, and the total fission width would be expected to have the one-channel distribution $x^{-\frac{1}{2}} \exp(-\frac{1}{2}x)$.

The distributions of nuclear reaction widths enter into the considerations of the averages and the fluctuations of the cross-sections for nuclear reactions which proceed through compound-nucleus formation. Although in some of the previous published work the randomness of the signs of the reduced-width amplitudes was considered, it was generally not suspected that the fluctuations of the magnitudes were large enough to warrant detailed considerations.[13] The average cross-section for compound elastic scattering of neutrons, for example, is now found to depend critically on the extent of the fluctuations of the neutron reduced widths. Thus, the cross-section predicted using the $x^{-\frac{1}{2}} \exp(-\frac{1}{2}x)$ distribution is twice as large as that predicted using the less-violent $\exp(-x)$ distribution, and in the excitation region where the levels overlap it is many times larger than the prediction for constant widths.[14] As another example, the average capture-to-fission ratio of U^{235} is found to exceed the ratio of the average capture width to the average fission width by an amount which depends on the extent of the fission width fluctuations.[15] These matters will be discussed later in more detail.

II. Neutron Widths

The distributions $x^{-\frac{1}{2}} \exp(-\frac{1}{2}x)$ and $\exp(-x)$ belong to the class of chi-squared distributions

$$P(x;\rho)dx = \Gamma(\rho)^{-1} (\rho x)^{\rho-1} e^{-\rho x} \rho dx, \tag{1}$$

where $\nu = 2\rho$ is a parameter which is referred to in the literature on statistics as the number of degrees of freedom. This terminology is also appropriate for physical discussions, and the distributions under investigation will be referred to as chi-squared distributions with one and two degrees of freedom, respectively. The above expression is recognized as being proportional to the integrand of the integral which defines the gamma function; the gamma function $\Gamma(\rho)$ in (1) thus serves to normalize $P(x)$ to unity. It is evident that $\langle x \rangle_{\text{Av}} = 1$ and that the variance of $x(\text{var} x \equiv \langle x^2 \rangle_{\text{Av}} - \langle x \rangle_{\text{Av}}^2)$ is equal to ρ^{-1} so that the parameter ρ (or, equivalently, the number of degrees of freedom ν) characterizes the width of the distribution, the greater the number the narrower the distribution. Chi-squared distributions for several integral values of ν are drawn

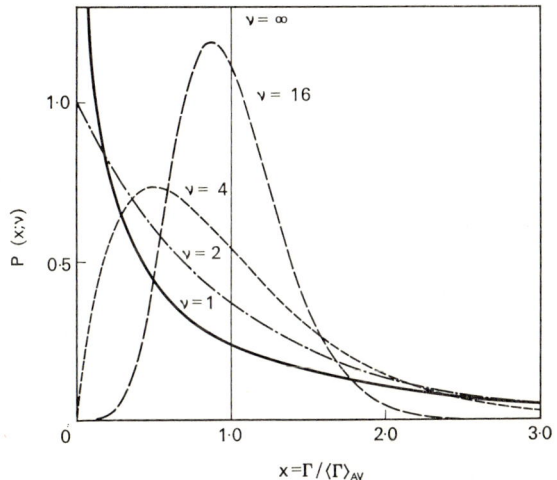

FIG. 1. The chi-squared distribution (1) is plotted for $\nu = 1, 2, 4, 16$, and ∞ degrees of freedom. The abscissa x is the ratio of the width Γ to its average $\langle \Gamma \rangle_{\text{Av}}$. Note that $\nu = 2\rho$.

in Fig. 1: when $\rho \geqslant 1$, the maxima (most-probable values) appear at $x = 1 - \rho^{-1}$; when $\rho \leqslant 1$, the function becomes infinite at the origin; and when $\rho = \infty$, it reduces to a delta function at $x = 1$. The chi-squared distribution can thus describe a wide variety of distributions, and the object is now to determine the range of ν (considered as a continuous variable) that is reasonably consistent with the data and to test the hypotheses that the "true" distribution has one or that it has two degrees of freedom.[16]

A statistically efficient method (that is, one that admits a small uncertainty) for determining the best-fitting value of the parameter ρ is the maximum-likelihood method.[17] According to this method the most-probable value of ρ is the one that maximizes the logarithm of the likelihood function, which is the product of the $P(x_i; \rho)$ for the set of m measurements x_i. In this way it is found that the most probable value of ρ is the one that satisfies the

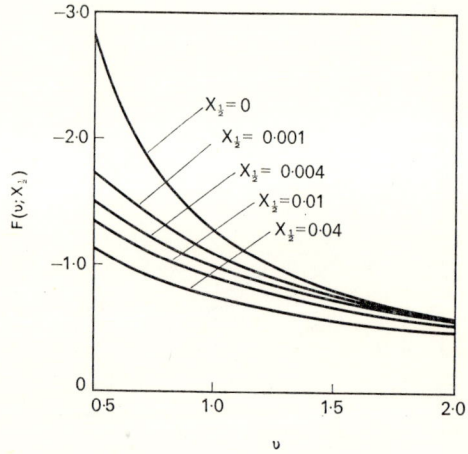

FIG. 2. Plot of the function $F(\rho; x_{\frac{1}{2}})$ of (11) for several values of the parameter $x_{\frac{1}{2}}$, the value of x at which the efficiency for detecting levels is one-half. The most probable value of $x_{\frac{1}{2}}$ is 0.01. When $x_{\frac{1}{2}} = 0$ this function reduces to the function defined by (2a). Note that $\nu = 2\rho$.

transcendental equation (the subscript and superscript of $\Gamma_n{}^0$ being dropped)

$$\frac{1}{m} \sum_{i=1}^{m} \ln(\Gamma_i / \langle \Gamma \rangle_{\mathrm{Av}}) - F(\rho) = 0, \qquad (2)$$

where

$$F(\rho) = \psi(\rho) - \ln \rho, \qquad (2a)$$

$\psi(\rho)$ being the derivative of the logarithm of the gamma function. The curve in Fig. 2 labeled $x_{\frac{1}{2}} = 0$ is the function $F(\rho)$, and it may be used for the determination of ρ when the sum in (2) is known. The asymptotic expression for the variance of this determination is

$$\text{var}_s \rho = m^{-1}[\psi'(\rho) - \rho^{-1}]^{-1}, \tag{2b}$$

where ψ' is the derivative of ψ.[18]

For U^{238} there are eleven reduced-width measurements that can be used.[19] By approximating the population average $\langle \Gamma \rangle_{\text{Av}}$ by the sample average,

$$\bar{\Gamma} = \frac{1}{m} \sum_{i=1}^{m} \Gamma_i, \tag{3}$$

a value of -1.22 is obtained for the sum in (2), and referring to the $x_{\frac{1}{2}} = 0$ curve of Fig. 2 a value of $\nu = 1.04$ degrees of freedom is deduced. According to (2b) the standard deviation of this determination is 0.36 degree of freedom, indicating that the value $\nu = 1$ is consistent with the data whereas the value $\nu = 2$ is not.

Although the above example does illustrate the main features of the maximum likelihood analysis, several important refinements are called for: 1. The sample sizes available for the various nuclides are not large, and a correction must be made for a possible difference of the sample average from the population average. 2. Samples of reasonable size are available for about 15 nuclides, and an analysis is desired which makes use of all of these for the estimate of the best universal value of ν. 3. The experimental uncertainties of the reported reduced widths should be taken into account. 4. As a consequence of instrumental limitations, levels with small widths may escape detection, and although there may be only a few of these, rather large errors can be introduced into the ν determination, especially when ν is small, as it seems to be. These complications will now be considered.

1. The sample average (3) has a chi-squared distribution with $m\nu$ degrees of freedom[20]:

$$P(\bar{\Gamma}; \rho)d\bar{\Gamma} = \frac{1}{\Gamma(m\rho)} \left(\frac{m\rho\bar{\Gamma}}{\langle\Gamma\rangle_{\text{Av}}}\right)^{m\rho-1} \times \exp\left(-\frac{m\rho\bar{\Gamma}}{\langle\Gamma\rangle_{\text{Av}}}\right) d\left(\frac{m\rho\bar{\Gamma}}{\langle\Gamma\rangle_{\text{Av}}}\right). \tag{4}$$

From this distribution it is evident that

$$\langle \bar{\Gamma} \rangle_{Av} = \langle \Gamma \rangle_{Av}, \tag{4a}$$

$$\text{var}_s \bar{\Gamma} = \langle \Gamma \rangle_{Av}^2/m\rho, \tag{4b}$$

$$\langle \ln(\bar{\Gamma}/\langle \Gamma \rangle_{Av}) \rangle = F(m\rho), \tag{4c}$$

and that the most probable value of $\bar{\Gamma}$ is

$$[1 - (m\rho)^{-1}] \times \langle \Gamma \rangle_{Av}.$$

Consider the function

$$\Phi = \frac{1}{m} \sum_{i=1}^{m} \ln(\Gamma_i/\bar{\Gamma}) + F(m\rho) - F(\rho). \tag{5}$$

Using (4c) it is evident that

$$\langle \Phi \rangle_{Av} = 0, \tag{5a}$$

and it may be shown using (1) and (4) that

$$\text{var}_s \Phi = m^{-1} \psi'(\rho) - \psi'(m\rho). \tag{5b}$$

The best estimate for ρ may be taken as the solution to

$$\Phi(\rho) = 0. \tag{5c}$$

According to the central-limit theorem, as the sample size increases the distribution of the function Φ asymptotically approaches normality:

$$P(\Phi)d\Phi \sim (2\pi \text{ var}\Phi)^{-\frac{1}{2}} \exp(-\Phi^2/2 \text{ var}\Phi)d\Phi. \tag{5d}$$

It will be assumed that the samples are large enough to justify a normal approximation (5d), and the hypotheses may then be tested by consideration of the error function

$$\mathscr{P} = \left(\frac{2}{\pi}\right)^{\frac{1}{2}} \int_{|\alpha|}^{\infty} \exp(-\tfrac{1}{2}z^2)dz, \tag{5e}$$

where

$$\alpha = \Phi/(\text{var}\Phi)^{\frac{1}{2}}.$$

It is evident from (4b) that the variance of an average determination of a sample of widths which obey a $\nu = 1$ distribution is twice as large as it is for a $\nu = 2$ distribution. In many of the published listings of average reduced widths and of strength functions, the uncertainties were arrived at assuming $\nu = 2$; these uncertainties should be increased accordingly if the $\nu = 1$ distribution is accepted as the correct one.[21] For the U^{238} data the average of the 11 widths is 2·7 mV (1 mV $\equiv 10^{-3}$ eV), and the standard deviation corresponding to (4b) is 1.2 mV, indicating that the errors of the average estimates from the typical samples available for heavy elements are apt to be rather large.

2. It is a straightforward matter to modify the above formulas for treatment of the composite data. In place of (5), one considers the function

$$\Phi = (1/m) \Sigma_{ij} \ln(\Gamma_{ij}/\bar{\Gamma}_j) + \Sigma_j(m_j/m)F(m_j\rho) - F(\rho), \quad (6)$$

where Γ_{ij} is the ith width of the jth nuclide; m_j is the number of the widths of the jth nuclide; and $m = \Sigma_j m_j$. It may be verified that

$$\langle \Phi \rangle_{Av} = 0, \quad (6a)$$

and that

$$\text{var}_s \Phi = (1/m)\psi'(\rho) - \Sigma_j(m_j/m)^2 \psi'(m_j\rho). \quad (6b)$$

For even the smallest sample which is used ($m_j = 3$), it is sufficiently accurate to use for $F(m_j\rho)$ and for $\psi'(m_j\rho)$ just the first few terms of the asymptotic expansions[22]:

$$\psi(z) - \ln z \sim -\tfrac{1}{2}z^{-1} - (1/12)z^{-2} + (1/120)z^{-4} - \cdots, \quad (7a)$$

$$\psi'(z) \sim z^{-1} + \tfrac{1}{2}z^{-2} + \tfrac{1}{6}z^{-3} - (1/30)z^{-5} + \cdots. \quad (7b)$$

3. If the possible values of the errors of the Γ_{ij} in the first sum in (6) are assumed to be small, independent, and normally distributed with zero means and with standard deviations σ_{ij}, the additional contribution to the variance of Φ is found to be

$$\text{var}_e \Phi = m^{-2} \Sigma_{ij}[\sigma_{ij}(\Gamma_{ij} - \bar{\Gamma}_j)/\Gamma_{ij}\bar{\Gamma}_j]^2. \quad (8)$$

The additional contribution to the variance (4b) of $\bar{\Gamma}_j$ is

$$\text{var}_e \bar{\Gamma}_j = m_j^{-2} \sum_{i=1}^{m_j} \sigma_{ij}^2. \quad (9)$$

For the evaluation of (5e), the contributions (5b) and (8) may be added as though they were independent.

4. Since the instrumental resolution depends upon the neutron energy (as does the average neutron width), the efficiency for detecting levels will also depend on the neutron energy as well as on the average reduced width of the sample. An analysis which accounts for these dependences has not been devised, and we will have to be content with using an over-all efficiency factor $E(x)$ for all energies and for all samples. The "experimental" distribution function will be taken as the product of the "theoretical" factor, which is given by (1), and the "experimental" factor $E(x)$, this product being then renormalized to unity. According to an analysis by Harvey,[23] the over-all efficiency is one-half at a value $x \equiv x_{\frac{1}{2}} =$ antilog$_{10}$ ($- 2.00 \pm 0.15$); the efficiency increases from 30 to 70 per cent for a factor 10 increase in x and from 15 to 85 per cent for a factor 100 increase in x; it is uncertain outside of the latter interval. As a function of $\log_{10}x$, the efficiency curve may be approximated closely by an error function with a standard deviation of 0.85. However, for the present considerations it is more convenient to represent it by a function of the form

$$E(x; x_{\frac{1}{2}}; \sigma) = 1 - \exp[-(\kappa x)^\sigma], \qquad (10)$$

where

$$\kappa = (\ln 2)^{1/\sigma}/x_{\frac{1}{2}},$$

and a good fit is obtained with a value of 0.53 for the shape parameter σ. The analysis has been carried out with $\sigma = \frac{1}{2}$, 1, and ∞.[24] The results were found to be essentially the same in each case so that it will suffice to present here the details for just the simplest case, $\sigma = 1$.

The quantity κ is very large compared with ρ, and it is therefore permissible to neglect $-\rho x$ when it appears together with $-\kappa x$ as the argument of the exponential function; after renormalization the "experimental" distribution is then

$$P(x)dx = (1 + q)\Gamma(\rho)^{-1}(\rho x)^{\rho-1}(e^{-\rho x} - e^{-\kappa x})\rho dx, \qquad (1')$$

where

$$q = (\zeta^\rho - 1)^{-1}, \quad \zeta = \kappa/\rho.$$

We were unable to derive from (1') the distribution corresponding to (4) for the sample average. However, the low-order moments can readily be derived from (1'); to a high degree of accuracy one finds that

$$\langle \bar{\Gamma} \rangle_{Av} = (1 + q)\langle \Gamma \rangle_{Av}, \tag{4a'}$$

$$\text{var}_s \bar{\Gamma} = \langle \bar{\Gamma} \rangle_{Av}^2 / \bar{m}\rho, \tag{4b'}$$

where

$$\bar{m} = m(1 + q)(1 - \rho q)^{-1}.$$

For the final analysis it is sufficiently accurate to approximate the distribution for the sample average by (4) with $\langle \Gamma \rangle_{Av}$ replaced by $(1 + q)\langle \Gamma \rangle_{Av}$ and with m replaced by $\bar{m}$, so that the correct expectation values (4a') and (4b') are obtained. With this approximation, one can write

$$\langle \ln[\bar{\Gamma}/(1 + q)\langle \Gamma \rangle_{Av}] \rangle_{Av} = F(\bar{m}\rho), \tag{4c'}$$

and it may be shown that (6a) is very nearly satisfied if in (6) the function $F(m_j\rho)$ is replaced by $F(m_j\rho)$ and if the function $F(\rho)$ is replaced by

$$F(\rho; x_{\frac{1}{2}}) = F(\rho) + q \ln \zeta - \ln(1 + q). \tag{11}$$

The expression corresponding to (6b) is found to be

$$\text{var}_s \Phi = (1/m)[\psi'(\rho) - q(1 + q)\ln^2\zeta] - \Sigma_j(m_j/m)^2 \psi'(\bar{m}_j\rho). \tag{6b'}$$

The function $F(\rho; x_{\frac{1}{2}})$ is plotted in Fig. 2 for several half-efficiency values $x_{\frac{1}{2}}$, including the most-probable value $x_{\frac{1}{2}} = 0.01$. It is evident that when $\nu = 1$ the function is rather sensitive to $x_{\frac{1}{2}}$, but when $\nu = 2$ it is not. With $x_{\frac{1}{2}} = 0.01$ and $\nu = 1$, the factor $(1 + q) = 1.09$, and (4a') indicates that to estimate $\langle \Gamma \rangle_{Av}$ the sample average should be reduced by about 9 per cent, which amount is usually small compared with the statistical uncertainty corresponding to (4b'). The quantity $\bar{m}$ in (4b') is equal to $1.15m$.

The expressions (5) through (9) with the modifications indicated under item 4 were used to analyse a total of 148 neutron reduced widths for 15 different nuclides. This total includes the 3 recent determinations for manganese[25] in addition to the 145 values which were analyzed by Hughes and Harvey.[4,26] The sum over ij in (6) is found to be -0.795 and the sum over j is -0.091 for $\nu = 1$ and -0.050 for $\nu = 2$; by using the $x_{\frac{1}{2}} = 0.01$ curve of Fig. 2, a value $\nu = 1.02$ is obtained as the solution to (5c). The standard deviations corresponding to (6b′) and (8) are 0.062 and 0.020, respectively, for $\nu = 1$, giving an over-all standard deviation of 0.065, which is primarily due to statistics. The standard deviation of the ν estimate corresponding to the combined variances of Φ is about 0.13 degree of freedom. The hypothesis $\nu = 1$ gives a probability integral $\mathscr{P}$, eq. (5e), which is close to unity, indicating that this hypothesis is quite consistent with the data, while the hypothesis $\nu = 2$ gives a value $a = -6.4$ and an inadmissibly small value of $\mathscr{P}$. The $\nu = 1$ hypothesis would be acceptable for any value of $x_{\frac{1}{2}}$ in the range antilog$_{10}$ (-2.00 ± 0.15) specified by Harvey.[23] However, for $x_{\frac{1}{2}} = 0$, $\mathscr{P}$ is only 0.006 for $\nu = 1$, thus indicating the importance of the efficiency correction; for $\nu = 2$ it would still be extremely small.

There is one datum which appears to be at variance with the $\nu = 1$ hypothesis. The plotted point for $y \equiv x_{\frac{1}{2}} = 0.1$ on Fig. 2 of the paper by Hughes and Harvey[4] falls way below the curve for $\nu = 1$ and near to the curve for $\nu = 2$. Since this point just corresponds to the most-probable half-efficiency value $x_{\frac{1}{2}} = 0.01$, it should be raised by a factor of about two, thus placing it almost within a standard deviation of the $\nu = 1$ curve and several standard deviations away from the $\nu = 2$ curve.

With the exception of U^{238}, all nuclides of footnote 26 can form two different spin states when interacting with low-energy neutrons. It has been assumed here that the two spin states have identical reduced-width distributions. At the time that this analysis was undertaken, this assumption seemed reasonable and was not known to be in contradiction with any data. However, from a recent summary of the data on the angular momentum associated with slow-neutron resonances, Sailor has found indications that the compound nucleus is preferentially formed in the state of higher spin.[27] If the two spin states are equally abundant, this indication implies that the average neutron width is not the same in each state. Unless the widths of the states of the lower spin are unobservably small, our analysis is apt to be biased towards a v value which is too small. With the existing data it is difficult to estimate the extent of this bias.

Before proceeding with the radiation and fission widths, it is worthwhile to include an explanatory remark on the physical significance of the terminology "degrees of freedom". In section I, arguments were put forth for the plausibility of a Gaussian amplitude distribution, corresponding to a chi-squared width distribution with one degree of freedom. These arguments took account of the fact that the amplitudes could suitably be chosen as real.[8] Now, if the amplitudes had been regarded as complex with independent real and imaginary parts, these arguments would have led to a width distribution with two degrees of freedom one for the real part and one for the imaginary part.[28] In the next section on radiation widths a situation is illustrated involving effectively a large number of degrees of freedom, and in the following section on fission widths one encounters a distribution with effectively only a few degrees of freedom.

III. Radiation Widths

The total radiation width Γ which pertains to the neutron resonance region is expected for heavy nuclei to be the sum of a large number n of partial radiation widths Γ_i:[29]

$$\Gamma = \sum_{i=1}^{n} \Gamma_i \qquad (12)$$

We will examine here the consequences of assuming that the distributions of the individual Γ_i are independent and chi-squared with one degree of freedom, like the neutron widths. As mentioned in section I, the latter assumption seems reasonable, while the former one stands on less certain grounds because some correlations are expected in the distributions of the partial widths for transitions to states of low excitation.

If all of the partial widths had the same average value $\langle \Gamma_i \rangle_{Av}$, the probability distribution for the total width would be chi squared with n degrees of freedom; that is, it would be given by (4) with $m\bar{\Gamma}$ replaced by Γ and with ρ replaced in $\frac{1}{2}n$. The distribution for the total width would thus become narrower as the number of partial widths becomes larger. For the general case where the $\langle \Gamma_i \rangle_{Av}$ are unequal, we have succeeded in deriving the distributions for the total only for even values of the number of degrees of freedom ν of the partial width distribution. However, these total distributions are rather complicated and all we need to consider anyway are the average and the variance; for any value of ν they are[30]

$$\langle \Gamma \rangle_{Av} = \sum_{i=1}^{n} \langle \Gamma_i \rangle_{Av}, \tag{13a}$$

$$\langle (\Gamma - \langle \Gamma \rangle_{Av})^2 \rangle_{Av} = (2/\nu) \sum_{i=1}^{n} \langle \Gamma_i \rangle_{Av}^2. \tag{13b}$$

These equations show that in the general case the distribution also becomes narrower as the number of partial widths becomes larger. In the following it is assumed that $\nu = 1$.

The partial radiation widths are believed to be proportional to E_γ^{2l+1}, where E_γ is the transition energy and l is its multipolarity.[29] By approximating the sum over radiative transitions by an integration using the level density formula

$$\rho(E) = C \exp(E/T), \tag{14}$$

with constant temperature T and a constant coefficient C, one finds that the average of any power m of E_γ is

$$\langle E_\gamma^m \rangle_{Av} = CT^{m+1} e^r \gamma(m+1, r), \tag{15}$$

where $r = B/T$, B being the neutron binding energy, and

$$\gamma(\alpha, x) \equiv \int_0^x t^{\alpha-1} e^{-t} dt$$

is an incomplete gamma function. For electric dipole radiation, the square of the *coefficient of variation* V is therefore predicted to be

$$V^2 \equiv \langle (\Gamma - \langle\Gamma\rangle_{Av})^2\rangle_{Av}/\langle\Gamma_{Av}\rangle_{Av}^2 = 2\langle E_\gamma^6\rangle_{Av}/\langle E_\gamma^3\rangle_{Av}^2$$
$$= 2n^{-1}(1 - e^{-r})[\gamma(7,r)/\gamma^2(4,r)], \qquad (16)$$

where

$$n = \int_0^B \rho(E)dE = CT(e^r - 1)$$

is the number of partial widths. Using Pearson's tables,[31] one finds that

$$V^2 = \frac{2}{n} \times \begin{cases} 2.3, r = 0 \\ 5.3, r = 3 \\ 10.9, r = 6 \\ 20.0, r = \infty. \end{cases} \qquad (17)$$

There are only a few elements for which a sufficient number of radiation widths have been determined for statistical considerations. The use of the above formulas will be illustrated by analyzing the widths for Ta^{181}, which has the largest reported number, although there may be some question as to their reliability; in mV they are: $49 \pm 6, 49 \pm 11, 50 \pm 10, 51 \pm 10, 50 \pm 15, 40 \pm 15$.[3] It is apparent that these widths show very little fluctuation other than that which could be ascribed to the indicated experimental uncertainties. The actual distribution must be narrow, and it may therefore be approximated by a normal distribution, the coefficient of variation of which may be compared with (17). As a generalization to the familiar chi-squared test of sample variances, it may be shown that the weighted sample variance has a chi-squared distribution with $m - 1$ degrees of freedom, where m is the number of measurements, the weighting factors being equal to the reciprocal of the sum of the population variance and the variance corresponding to the uncertainty of the measurement. Assuming to begin with that the population variance is zero [that is, that n in (17) is infinite], one arrives at a total probability of less than one per cent for weighted sample variances which are smaller than the observed one, indicating that there may be some systematic errors in the measurements or in the estimations of the uncertainties. For illustrative purposes, the uncertainties are neglected altogether, thus allowing one to state, for example, that there is only a 5 per cent total probability for the weighted sample variances being smaller than observed when a population variance of $(8.7 \text{ mV})^2$ is assumed; this variance corresponds to $n = 340$ in (17) with $r = 3$. This statement is consistent with the view that there are many independently-contributing partial widths, but it sheds very little light on the question of what fraction of all of the partial widths contribute independently. Thus, the number of partial widths is expected to be of the order of magnitude of the ratio of the nuclear temperature to the mean level spacing, which ratio is about 10^5 for a typical heavy nucleus.

The radiation widths of eleven nuclides for which more than one width are reported have carefully been analyzed for fluctuations by Levin and Hughes.[32] They found rather definite indications of fluctuations in the widths of several of the accurately measured nuclei, notably In^{115} and Eu^{151}.[33] However, for these two it is observed that the radiation widths fall into two groups, each group having a definite but distinct isomeric branching ratio, thus suggesting that each group corresponds to a different spin state of the compound nucleus. With this contingency, they concluded that it was not possible with the existing data to reject the hypothesis that the radiation widths of a particlar spin state of a particular nucleus are the same at all of the levels. A more critical testing of this hypothesis should be realized with a zero-spin target nucleus, with which only compound states of a single spin value could be excited with low-energy neutrons. The only such nucleus with a sufficient number of reported widths is U^{238}, these being 24 ± 2, 25 ± 5, 29 ± 9, 17 ± 10 mV.[3] From the chi-squared test of the weighted sample variance, it may be concluded that with a population variance of $(12 \text{ mV})^2$ there is only a 5 per cent probability for a sample variance smaller than the observed one. This population variance corresponds to a value of $n = 40$ in (17) with $r = 3$. This conclusion is not significantly different from the previous one for Ta^{181}.[34]

IV. Fission Widths (U^{235})

An analysis has been made of the fluctuations of the 15 fission widths of U^{235} which are provisionally reported in the recent compilation by Hughes and Egelstaff and in a private communication from Sailor.[35] These widths fluctuate considerably but not as much as do the neutron widths. The solution to eq. (5c) is $\nu = 2.3$ degrees of freedom, the standard deviations being 0.8 degree of freedom from the statistics and 0.3 degree of freedom from the indicated experimental uncertainties. The average width is 71 mV as estimated from the sample average, the standard deviations being 17 mV from the statistics and 5 mV from the indicated experimental uncertainties. To test the hypothesis that there is only one fission channel, that is, that the distribution has only one degree of freedom, the probability integral $\mathscr{P}(\nu = 1)$ of (5e) was evaluated and found to be 0.10, which is small but not inadmissible according to most statistical criteria.

Another way to estimate the number of degrees of freedom of the best-fitting chi-squared distribution is to equate the first and second moments of the sample to the corresponding moments of the population. In this way one estimates that[36]

$$\nu = 2(1 - m^{-1})/V^2, \qquad (18$$

where

$$V^2 = [m^{-1} \Sigma_i (\Gamma_i - \bar{\Gamma})^2]/\bar{\Gamma}^2,$$

the variance being given to order m^{-1} by

$$\text{var}_s \nu = \nu^2(1 + 4\nu^{-1})m/(m-1)^2. \qquad (18a)$$

Using the same data, these give 3.3 ± 1.6 degrees of freedom, which is consistent with the maximum likelihood estimate. The estimate (18) is especially poor when ν is small,[37] but for large ν the variances (2b) and (18a) may be shown, by using (7b), to be asymptotically equivalent.

The number of degrees of freedom will in general be smaller than the actual number of channels if the average widths for the various channels are unequal and if there are correlations in the distributions. Another difficulty in the interpretation is that there are two spin states formed when low-energy neutrons are captured by U^{235}, and these states will not necessarily have the same distributions.[38] The fact that the ratio of the average capture cross-section to the average fission cross-section at low energies[39] is about equal to the ratio of the average capture width[40] to the average fission width of the low-energy resonances indicates that the average fission widths for the two spin states are equal to within a factor of two. However, there is no way of telling from the existing data whether or not the distributions for the two spin states have the same variances. In spite of these complications, the original qualitative conclusions should remain valid: namely, that there are not very many channels involved in the slow-neutron-induced fission of U^{235}, and the likelihood is small of there being only one channel (for each spin state). The main reason for seeking the best fitting chi-squared distribution is that this distribution is very convenient to use for calculating the effects of the fission width fluctuations on the averages and variances of fission and competing reaction cross-sections.[41]

Interference effects have been noticed by Sailor[40] in the fission cross-section of U^{235}. It has also been noticed by him that the occurrence of interference effects in fission and the nonoccurrence of such effects in reactions in which radiative capture dominates are consistent with the view that there are many exit channels in the latter case, the reduced-width amplitudes of which have random signs, whereas there are effectively only a few such channels in the former case.

As a final remark, it is noted that there is no significant correlation in the fission and neutron width distributions of U^{235}. This observation confirms an earlier one by Harvey.[42] It also indicates that no correction to the analysis using (5c) is needed for failure to detect levels having very small neutron widths.[23]

Acknowledgements

The writers are grateful to J. A. Harvey for supplying the very important data on the efficiency for detecting weak resonances, to D. J. Hughes for a prepublication copy of the fission-width listing, to Max Goldstein for supervising the numerical computations, to V. L. Sailor for his comments on the reliability of the neutron resonance data and for a listing of U^{235} fission widths, and to R. K. Zeigler for suggesting the use of the maximum-likelihood method.

Notes

[1] Seidl, Hughes, Palevsky, Levin, Kato, and Sjöstrand, *Phys. Rev.* **95**, 476 (1954).

[2] Harvey, Hughes, Carter, and Pilcher, *Phys. Rev.* **99**, 10 (1955).

[3] D. J. Hughes and J. A. Harvey, *Neutron Cross Sections*, Brookhaven National Laboratory Report BNL-325 (Superintendent of Documents, U.S. Government Printing Office, Washington, D.C., 1955).

[4] D. J. Hughes and J. A. Harvey, *Phys. Rev.* **99**, 1032 (1955).

[5] The existence of a reasonably well-defined average neutron reduced width is assumed here. The existence of such an average is suggested by the work of Feshbach, Porter, and Weisskopf, *Phys. Rev.* **96**, 448 (1954) and of Lane, Thomas, and Wigner, *Phys. Rev.* **98**, 693 (1955).

[6] More precisely, the matrix element is a "surface" integral in the sense that the coordinate for the relative separation of the neutron and the residual nucleus is set equal to the channel radius. See eq. (17) of E. P. Wigner and L. Eisenbud, *Phys. Rev.* **72**, 29 (1947) for the precise definition.

[7] See, for example, Harald Cramér, *Mathematical Methods of Statistics* (Princeton University Press, Princeton, 1945), sec. 17.4.

[8] In this connection it is important to realize that the matrix elements can always be made real. See the discussion following the equation referred to in note 6.

[9] The relation between the reduced width γ^2 of the Wigner-Eisenbud theory (note 6) and the reduced width Γ_n^0 which is commonly used in the discussions of low-energy neutron resonance is $\Gamma_n^0 = \Gamma_n/E_0(\text{ev})^{\frac{1}{2}} = 0.4390\gamma^2 a \times 10^{-3}$, where γ^2 has the same energy unit as Γ_n^0, and a is the channel radius in units of 10^{-13} cm; although γ^2 refers to the center-of-mass system, the energies of Γ_n^0 and E_0 refer to the laboratory system. (As it is not necessary to specify a nuclear radius in the consideration of s-wave neutron widths, one may prefer to regard the reduced width as the product $\gamma^2 a$, which has the dimensions of energy times distance.) The upper limit of γ^2 is approximately $\hbar^2/Ma^2$, where M is the neutron mass, or $40/a^2$ MeV when a is specified in the above unit. The upper limit of Γ_n^0 is therefore $18/a$ keV.

[10] The proton reduced widths of the 2^+ levels of Cu^{59} which are excited when Ni^{58} is bombarded by protons with energies from $2\frac{1}{2}$ to 5 MeV are observed to be distributed in a exponential-like manner [J. P. Schiffer (private communication)]. For an abstract on this experiment, see Moore, Schiffer, and Class, *Bull. Am. Phys. Soc.* Ser. II. **1**, 39 (1956).

[11] Distributions which are independent are of course also uncorrelated, but the converse is not generally true. However, for the discussions which follow it is well to keep in mind that normal distributions which are uncorrelated are also independent (see sec. 24.1 of note 7).

[12] A. Bohr, *Proceedings of the International Conference on the Peaceful Uses of Atomic Energy* (Columbia University Press, New York, 1956), Vol. 2, Report P/911.

[13] See, for example, R. G. Thomas, *Phys. Rev.* **97**, 224 (1955).

[14] See eq. (45) of the reference quoted in in note 13.

[15] Sophie Oleksa (to be published).

[16] The distribution $x^{-\frac{1}{2}} \exp(-x^{\frac{1}{2}})$ was also tested by Hughes and Harvey,[4] and found to be inconsistent with the aggregate of the data. This distribution was observed by Bethe to give a good account of the reduced widths of U^{235} [H. A. Bethe, *Proceedings of the International Conference on the Peaceful Uses of Atomic Energy* (Columbia University Press, New York, 1956), Vol. 4, Report P/585]. We will not consider it here because the chi-squared distribution seems general enough, and for the same reason we will not consider any of a large variety of other distributions that might be proposed.

[17] See, for example, M. G. Kendall, *The Advanced Theory of Statistics* (Charles Griffin and Company, Ltd., London, 1946), Vol. II, chap. 17.

[18] For tables of $\Gamma(\ )$, ψ, and ψ' see Harold T. Davis, *Tables of Higher Mathematical Functions* (The Principia Press, Inc., Bloomington), Vol. I (1933) and Vol. II (1935).

[19] These are the measurements which appear above the dotted line in the table of resonance parameters of heavy nuclei in the reference quoted in note 3. The listing below the dotted line is not considered to be complete.

[20] See M. G. Kendall, *The Advanced Theory of Statistics* (Charles Griffin and Company, Ltd., London, 1946), Vol. I, Example 10.11. This result may readily be verified by noting that its Laplace transform is equal to the mth power of the Laplace transform of the distribution (1), in agreement with the convolution theorem.

[21] There are several indirect methods for determining the strength function which effectively make use of very large samples, and they are therefore not subject to this uncertainty: see S. E. Darden, *Phys. Rev.* **99**, 748 (1955); D. J. Hughes and V. E. Pilcher, *Phys. Rev.* **100**, 1249 (A) (1955).

[22] Erdélyi, Magnus, Oberhettinger, and Tricomi, *Higher Transcendental Functions* (McGraw-Hill Book Company, Inc., New York, 1953), Vol. I, sec. 1.18, eq. (7).

[23] J. A. Harvey (private communication).

[24] The analysis for $\sigma = \infty$ was actually carried out by truncating the distribution (1) at $x_{\frac{1}{2}}$ and by replacing the normalizing gamma function by the appropriate incomplete gamma function.

[25] Bollinger, Dahlbert, Palmer, and Thomas, *Phys. Rev.* **100**, 126 (1955).

[26] The actual values and their uncertainties were read from the listing of resonance parameters in the reference quoted in note 3. The nuclei were: $Mo^{95}(4)$; $Mo^{97}(4)$; $In^{113}(7)$; $In^{115}(8)$; $Sn^{117}(5)$; $Cs^{133}(12)$; $Eu^{151,153}(14)$; $Tb^{159}(16)$; $Ho^{165}(10)$; $Tm^{169}(10)$; $Hf^{177}(12)$; $Hf^{179}(17)$; $Ta^{181}(10)$; $U^{238}(11)$; the number of widths used for each nucleus is indicated in parentheses. Only those widths appearing above the dotted lines in the listings were used.

[27] V. L. Sailor (to be published).

[28] See J. M. C. Scott, *Phil. Mag.* **45**, 1322 (1954).

[29] See J. M. Blatt and V. F. Weisskopf, *Theoretical Nuclear Physics* (John Wiley and Sons, Inc., New York, 1952), chap. XII; and the article by B. B. Kinsey, in *Beta- and Gamma-Ray Spectroscopy* (Interscience Publishers, Inc., New York, 1955), pp. 795–822.

[30] See, for example, S. Chandrasekhar, *Revs. Modern Phys.* **15**, 1 (1943), appendix IV.

[31] K. Pearson, *Tables of the Incomplete Gamma Function* (Cambridge University Press, Cambridge, 1934).

[32] J. S. Levin and D. J. Hughes, *Phys. Rev.* **101**, 1328 (1956).

[33] H. H. Landon and V. L. Sailor, *Phys. Rev.* **98**, 1267 (1955). See also H. H. Landon, *Phys. Rev.* **100**, 1414 (1955); G. Igo and H. H. Landon, *Phys. Rev.* **101**, 726 (1956).

[34] Six radiation widths for U^{238} have recently been reported by J. E. Lynn and N. J. Pattenden, *Proceedings of the International Conference on the Peaceful Uses of Atomic Energy* (Columbia University Press, New York, 1956), Vol. 4, Report P/423. They are 26.1 ± 1.5, 28.8 ± 2.3, 24.9 ± 4.2, 18.6 ± 2.7, 15.5 ± 5.4, 13.6 ± 4.8 mV, thus revealing rather significant fluctuations. With these data, one could state with essentially 95 per cent confidence that n lies in the range from about 30 to 500.

[35] D. J. Hughes and P. A. Egelstaff, *Progress in Nuclear Energy* (Pergamon Press, London, 1956), Vol. 1, chap. II. The widths in the compilation were used except for those which have recently been determined more accurately by Sailor; the values and their uncertainties are: 99 ± 5, 120 ± 15, 13 ± 3, 110 ± 45, 130 ± 25, 87 ± 15, 3 ± 2, 9.5 ± 5, 10.5 ± 5, 70 ± 16, 42 ± 18, 43 ± 21, 90 ± 26, 200 ± 20, 44 ± 20 mV.

[36] See sec. 27.7 of the reference quoted in note 7, in particular eq. (27.7.3). The derivations of eqns. (18) and (18a) above are very similar to those of the example presented in connection with eqs. (27.7.10) of this reference. However, here we treat the square of V whereas the example treats V. The expectation value of V^2 has also been evaluated to one higher order in m^{-1} in order to obtain the factor $(1 - m^{-1})$ of (18) which makes the estimate unbiased.

[37] See sec. 17.51 of the reference quoted in note 17.

[38] The spin of U^{235} is 7/2 [K. L. Vander Sluis and J. R. McNally, Jr., *J. Opt. Soc. Am.* **45**, 65 (1955)] and its parity is presumably even [M. G. Mayer and J. H. D. Jensen, *Elementary Theory of Nuclear Shell Structure* (John Wiley and Sons, Inc., New York, 1955), p. 81] so that even-parity states of spin 3 and 4 are formed. From the theoretical account of slow-neutron-induced fission which is given in the reference quoted in note 12, it is implied that the average fission width of the 4^+ states will be larger than that of the 3^+ states, because the former type state is contained in the rotational band associated with the

lowest nucleonic configuration while the latter type would appear in a band involving an excited configuration. This configuration excitation is estimated to be of the order of a MeV, so that the difference of the average fission widths is expected to be especially large if the slow-neutron excitation energy exceeds the absolute threshold for fission of the 4+ states by less than about one MeV. However, recent results on the U^{235} (*d*, *p*) *fission* reaction indicate that this excess is about $1\frac{1}{2}$ MeV (unpublished experiments of K. Boyer and R. H. Stokes), and it is perhaps not inconsistent with the theory for the difference of the widths to be small.

[39] Kanne, Stewart, and White, *Proceedings of the International Conference on the Peaceful Uses of Atomic Energy* (Columbia University Press, New York, 1956), Vol. 4, Report P/595.

[40] V. L. Sailor, *Proceedings of the International Conference on the Peaceful Uses of Atomic Energy* (Columbia University Press, New York, 1956), Vol. 4, Report P/586.

[41] A likelihood analysis of the 15 fission widths for Pu^{239}, which are provisionally reported by L. M. Bollinger, R. E. Coté, J. M. LeBlanc, and G. E. Thomas, gives for the number of degrees of freedom $\nu \lesssim 1.7$ with an uncertainty of 0.5 degree of freedom from the statistics. Only an upper limit to ν can be stated because one of the resonances has no detectable fission width.

[42] J. A. Harvey, *Bull. Am. Phys. Soc.* Ser. II, **1**, 86 (1956).

Appendix 1

The relation between the descriptions of the fission process in terms of fission channels and in terms of fission fragment pairs may be illustrated by a few simple equations. The wave functions ψ_c for the fission channels c may be expanded in terms of the wave functions φ_p for the fission fragment pairs p as

$$\psi_c = \Sigma_p a_{cp} \varphi_p \tag{19}$$

with real coefficients a_{cp}. Both sets of wave functions are assumed to be orthonormal (but not necessarily complete) so that

$$\Sigma_p a_{cp} a_{c'p} = \delta_{cc'} \tag{20}$$

On the "surface" $\mathscr{S}$ of the nuclear configuration space, the wave functions X_λ of the levels λ of the compound nucleus may be expressed in terms of the fission-channel states as

$$X_\lambda(\mathscr{S}) \sim \Sigma_c \gamma_{\lambda c} \psi_c \tag{21a}$$

plus terms associated with other reactions, where the coefficients $\gamma_{\lambda c}$ are the reduced-width amplitudes for the fission channels c. Using (19) the surface expansion may alternatively be expressed as

$$X_\lambda(\mathscr{S}) \sim \Sigma_p \gamma_{\lambda p} \varphi_p, \tag{21b}$$

plus the other terms, where

$$\gamma_{\lambda p} = \Sigma_c \gamma_{\lambda c} a_{cp} \tag{22}$$

is the reduced-width amplitude associated with the pth fission fragment pair. The total fission width $\Gamma_{\lambda f}$ for the level λ, which is normally expressed as the sum of partial widths for the various fragment pairs, may also be expressed as a sum over the partial widths associated with the various fission channels; using (20), one finds that

$$\Gamma_{\lambda f} = \Sigma_p \gamma_{\lambda p}{}^2 = \Sigma_c \gamma_{\lambda c}{}^2. \tag{23}$$

Now if the distributions of the $\gamma_{\lambda c}$ are normal with respect to levels, with zero means, and independent with respect to channels, then the distributions of the $\gamma_{\lambda p}$, as expressed by (22), will also be normal with variances

$$\langle \gamma_{\lambda p}{}^2 \rangle_{\text{Av}} = \Sigma_c a_{cp}{}^2 \langle \gamma_{\lambda c}{}^2 \rangle_{\text{Av}}, \tag{24a}$$

he averages being with respect to the levels, and with covariances (correlation coefficients)

$$\langle \gamma_{\lambda p} \gamma_{\lambda p'} \rangle_{\text{Av}} = \Sigma_c a_{cp} a_{cp'} \langle \gamma_{\lambda c}{}^2 \rangle_{\text{Av}}. \tag{24b}$$

Near the fission threshold it is expected that only a few fission channels will have appreciable amplitudes $\gamma_{\lambda c}$ so that the covariances will not generally vanish even though the φ_c may constitute a complete set. This means that the distributions of the $\gamma_{\lambda p}$ are expected to be correlated, and therefore to be dependent.[11]

6

The Scattering of High-energy Neutrons by Nuclei†

S. Fernbach, R. Serber and T. B. Taylor

> **Abstract.** The experiments of Cook, McMillan, Peterson, and Sewell on the cross-sections of nuclei for neutrons of about 90 MeV indicate that the nuclei are partially transparent to high-energy neutrons. It is shown that the results can be explained quite satisfactorily using a nuclear radius $R = 1.37 A^{\frac{1}{3}} \times 10^{-13}$ cm, a potential energy for the neutron in the nucleus of 31 MeV, and a mean free path for the neutron in nuclear matter of 4.5×10^{-13} cm. This mean free path agrees with that estimated from the high energy n-p cross-section, but the results are not sensitive to the choice of mean free path.

In a previous paper by one of the writers[1] it has been pointed out that to a high-energy bombarding particle a nucleus appears partially transparent, since at energies of the order of 100 MeV the scattering mean free path for a neutron or proton traversing nuclear matter becomes comparable to the nuclear radius. This transparency effect is strikingly apparent in the experiments of Cook, McMillan, Peterson, and Sewell[2] on the scattering by nuclei of neutrons of about 90 MeV. In the present paper it will be shown that the observed scattering cross-sections can be quite satisfactorily accounted for, using a mean free path of the expected magnitude.

The problem is that of the scattering of the neutron wave by a sphere of material characterized by an absorption coefficient and an index of refraction. The index of refraction is determined by the mean potential energy, V, of the neutron in the nucleus. If $k =$

† *Phys. Rev.* **75**, 1352 (1949).

$(2ME)^{\frac{1}{2}}/\hbar$ is the propagation vector of the wave outside the nucleus, its propagation vector inside is $k + k_1$, with

$$k_1 = k[(1 + V/E)^{\frac{1}{2}} - 1].$$

For $E = 90$ MeV, $k = 2.08 \times 10^{13}$ cm^{-1}. The potential V is generally taken to be about 8 MeV larger than the energy of the Fermi sphere. The latter depends on the assumed nuclear density. If we use for the nuclear radius the value $R = 1.37A^{\frac{1}{3}} \times 10^{-13}$ cm, deduced by Cook, McMillan, Peterson, and Sewell from the 14–25 MeV scattering results of Amaldi, Bocciarelli, Cacciapuoti, and Trabacchi,[3] and Sherr,[4] we find a Fermi energy of 22 MeV, and $V = 30$ MeV. This gives $k_1 = 3.22 \times 10^{12}$ cm^{-1}. The absorption coefficient in nuclear matter is equal to the particle density times the cross-section for scattering of the neutron by a particle in the nucleus,

$$K = 3A\sigma/4\pi R^3.$$

In terms of the *n-p* and *n-n* cross-sections,

$$\sigma = [Z\sigma_{np} + (A - Z)\sigma_{nn}]/A.$$

Cook *et al.*[2] give for the scattering of a 90 MeV neutron by a free proton $\sigma_{np(\text{free})} = 8.3 \times 10^{-26}$ cm^2. This cross-section must be reduced to allow for the effect of the exclusion principle on the scattering by a proton bound in the nucleus; according to Goldberger,[5] the factor is $\sigma_{np} = \frac{2}{3}\sigma_{np(\text{free})}$. Assuming a $1/E$ dependence of the cross-sections we find, for $E = 90 + 30 = 120$ MeV, $\sigma_{np} = 4.15 \times 10^{-26}$ cm^2. If, following Goldberger, we take $\sigma_{nn} = \frac{1}{4}\sigma_{np}$, and use the previously quoted radius formula, we obtain $K = 2.4 \times 10^{12}$ cm^{-1} for $Z/A = \frac{1}{2}$, $K = 2.1 \times 10^{12}$ cm^{-1} for $Z/A = 0.39$ (U). It will be seen from these numbers that in the ensuing calculations it will be a reasonable approximation to suppose that $kR \gg 1$, but k_1/k and $K/k \ll 1$, so that k_1R and KR are of order one.

The scattering cross-section consists of two parts. The first, the "absorption cross-section", is just πR^2 times the probability that the neutron collides with a particle in the nucleus. This is not true absorption: inelastic scattering and scattering with exchange are

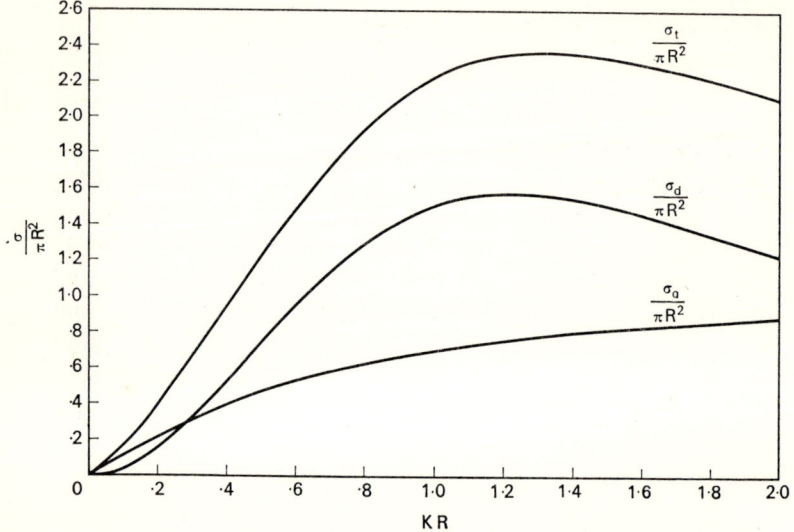

Fig. 1. Absorption, diffraction and total cross-section as a function of the nuclear radius measured in mean free paths. These curves are for $k_1/K = 1.5$.

included. The second part, the "diffraction scattering", is elastic scattering arising from the disturbance of the incident plane wave by the nucleus. To illustrate the calculation, we first consider the scattering from a disk of radius R and thickness T. We suppose there is a boundary layer at the surface of the disk in which k_1 and K rise to their interior values in a distance larger than $1/k$.[6] There will then be no scattering at the surfaces, and, for unit amplitude of incident wave, the wave transmitted through the disk will have an amplitude and relative phase $a = \exp(-\tfrac{1}{2}K + ik_1)T$. The absorption cross-section is

$$\sigma_a = \pi R^2(1 - |a|^2) = \pi R^2(1 - e^{-KT}). \tag{1}$$

The diffraction cross-section can be found from the consideration that on a plane behind the disk the wave is no longer plane, but

SCATTERING OF HIGH-ENERGY NEUTRONS BY NUCLEI

differs from a plane wave by an amplitude $1 - a$ in the shadow of the disk. This amplitude represents a scattered wave, and the corresponding cross-section is

$$\sigma_d = \pi R^2 \mid 1 - a \mid^2 = \pi R^2(1 - 2e^{-\frac{1}{2}KT}\cos k_1 T + e^{-KT}). \quad (2)$$

It can easily be shown that the angular dependence of the scattered amplitude is

$$f(\theta) = k \int_0^R (1 - a) J_0(k\rho \sin\theta) \rho d\rho$$
$$= (1 - a) R J_1(kR \sin\theta)/\sin\theta, \quad (3)$$

which gives the differential scattering cross-section

$$d\sigma_d(\theta) = |f(\theta)|^2 d\Omega = (\sigma_d/\pi)[J_1(kR\sin\theta)/\sin\theta]^2 d\Omega. \quad (4)$$

The absorption cross-section is, of course, always less than πR^2, but the diffraction cross-section may be either larger or smaller, depending on the magnitude of the phase shift. For large KT, $\sigma_a = \sigma_d = \pi R^2$. In the opposite limit of small KT and $k_1 T$, we have

$$\sigma_a = \pi R^2 KT = A\sigma,$$
$$\sigma_d = \pi R^2(\tfrac{1}{4}K^2 + k_1^2)T^2 = \tfrac{1}{4}A\sigma[1 + 4(k_1^2/K^2)]KT.$$

Thus for low density or small thickness, σ_a approaches the sum of the scattering cross-sections of the separate nucleons. The diffraction cross-section, however, vanishes in the limit, being proportional to the probability of double scattering.

The corresponding calculations for a sphere are only slightly more complicated. The portion of the wave which strikes the sphere at a distance ρ from a line through the center of the sphere emerges after traveling a distance $2s$, with $s^2 = R^2 - \rho^2$. Its amplitude on emerging is $a = \exp(-K + 2ik_1)s$, so that, in place of (1) we have

$$\sigma_a = 2\pi \int_0^R (1 - e^{-2Ks}) \rho d\rho = 2\pi \int_0^R (1 - e^{-2Ks}) s ds$$
$$= \pi R^2\{1 - [1 - (1 + 2KR)e^{-2KR}]/2K^2R^2\}. \quad (5)$$

This formula for the absorption cross-section has previously been given by Bethe.[7] Similarly, in place of (2), we have

$$\sigma_d = 2\pi \int_0^R |1 - e^{(-K+2ik_1)s}|^2 \, p\,dp$$

$$= \pi R^2[1 + (1/2K^2R^2)\{1 - (1 + 2KR)e^{-2KR}\}$$
$$- (1/(\tfrac{1}{4}K^2 + k_1{}^2)^2 R^2)\{(\tfrac{1}{4}K^2 - k_1{}^2)$$
$$+ e^{-KR}[2k_1R(\tfrac{1}{4}K^2 + k_1{}^2) + k_1K]\sin 2k_1R$$
$$- e^{-KR}[(\tfrac{1}{4}K^2 - k_1{}^2) + KR(\tfrac{1}{4}K^2 + k_1{}^2)]$$
$$\times \cos 2k_1R\}]. \qquad (6)$$

In deriving (5) and (6) we have neglected refraction at the surface of the sphere. It can easily be seen that this is legitimate, since it gives an effect of order $(k_1/k)k_1R$.

For the angular distribution we find, in analogy to (3),

$$f(\theta) = k\int_0^R [1 - e^{(-K+2ik_1)s}]J_0(k\rho \sin\theta) \, \rho \, d\rho. \qquad (7)$$

For $KR \to \infty$, we again obtain (4), but we have not found a convenient expression in the general case. The amplitude for forward scattering is easily evaluated, and is found to be

$$f(0) = \frac{kR^2}{2}\left\{1 + \frac{(k_1 - \tfrac{1}{2}iK)^2[1 - (1 + KR - 2ik_1R)e^{(-K+2ik_1)R}]}{2(\tfrac{1}{4}K^2 + k_1{}^2)^2 R^2}\right\}. \qquad (8)$$

For purposes of calculation, the integral can be converted to a sum; letting $l + \tfrac{1}{2} = k\rho$ and using the relation $J_0((l + \tfrac{1}{2})\sin\theta) = P_l(\cos\theta)$, valid for large l and small θ, we find

$$f(\theta) = \tfrac{1}{2}k \sum_{l=0}^{l+\tfrac{1}{2} < kR} (2l + 1)(1 - e^{(-K+2ik_1)s_l})P_l(\cos\theta), \qquad (9)$$

where

$$s_l = [k^2R^2 - (l + \tfrac{1}{2})^2]^{\tfrac{1}{2}}/k.$$

This expression can also be obtained by a partial wave analysis, using the WKB method to evaluate the phase shifts. This gives

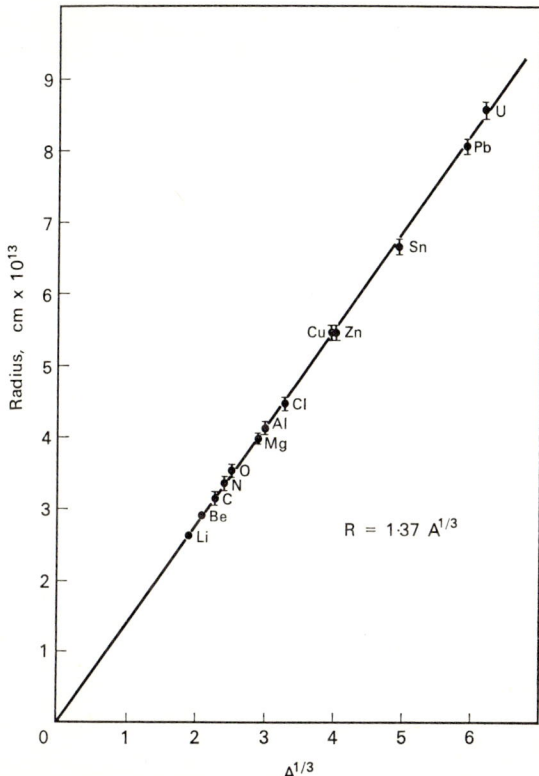

FIG. 2. Nuclear radii deduced from the total cross-section measurements of Cook, McMillan, Peterson, and Sewell plotted against the cube roots of the mass numbers. The straight line is $R = 1.37 A^{\frac{1}{3}} \times 10^{-13}$ cm.

$$\delta_l = (k_1 + \tfrac{1}{2}iK)s_l,$$

whence we immediately obtain (9), and for σ_a and σ_d,

$$\sigma_a = (\pi/k^2)\Sigma_l(2l+1)(1 - e^{-2Ks_l}), \tag{10}$$
$$\sigma_d = (\pi/k^2)\Sigma_l(2l+1)\,|\,1 - e^{(-K+2ik_1)s}\,|^2. \tag{11}$$

Converting the sums in (10) and (11) to integrals we again obtain (5) and (6).

In Fig. 1 we have plotted $\sigma_a/\pi R^2$, $\sigma_d/\pi R^2$, and the total cross-section $\sigma_t/\pi R^2 = (\sigma_a + \sigma_d)/\pi R^2$ as functions of KR. The ratio $\sigma_a/\pi R^2$ is a function only of KR; the other two depend on k_1/K as well. The curves in Fig. 1 have been plotted for $k_1/K = 1.5$, about the ratio indicated by our earlier consideration of the expected magnitude of the constants. Using this plot it is possible to determine, once a value of K is chosen, the radius required for each nucleus to give the measured total cross section. The radii calculated in this way from the observed cross-sections, using the value[8] $K = 2.2 \times 10^{12}$ cm^{-1}, are shown in Fig. 2. It will be seen that they lie quite closely on the line $R = 1.37 A^{\frac{1}{3}} \times 10^{-13}$ cm; the self-consistency of our description of the scattering process is thus established. The value $K = 2.2 \times 10^{12}$ cm^{-1} corresponds to a mean free path in nuclear matter of 4.5×10^{-13} cm. The associated value, $k_1 = 3.3 \times 10^{12}$ cm^{-1}, corresponds to $V = 30.8$ MeV.

The question now arises as to the accuracy with which the constants K and k_1 are determined by the scattering data. If k_1 is decreased, keeping K constant, it is found that the radius curve, Fig. 2, is pulled up in the middle; the resultant curve can be approximated by two straight lines, the light elements lying on a steeper line through the origin, while the heavy elements lie on a less steep line with a positive intercept. Increasing k_1 has the opposite effect. A variation in k_1 of $\pm 0.2 \times 10^{12}$ cm^{-1}, or in V or ± 2 MeV begins to produce appreciable bending. A reduction in K, with fixed k_1, introduces a curvature in the radius line, the center being pulled down and the two ends raised. The curvature becomes noticeable if K is reduced to less than $K = 1.9 \times 10^{12}$ cm^{-1}, however K can be almost doubled before the opposite curvature becomes very pronounced. For example, $K = 3.0 \times 10^{12}$ cm^{-1} gives an about equally good straight line, $R = 1.39 A^{\frac{1}{3}} \times 10^{-13}$ cm. The total cross-section measurements thus determine the potential fairly well, but are quite insensitive to the absorption coefficient. Measurements of σ_a and of the differential diffraction scattering are required for a better evaluation of K. It should be noted that while k_1 and K are determined directly from the cross-sections, the evaluation of V depends also on the energy of the incident neutrons. Cook *et al.* state that the energy of the neutrons detected in their experiment may be a little lower than 90 MeV, lying somewhere between 80 and 90 MeV. If we took $E = 80$ MeV, we would find $V = 28.8$ MeV.

For $K = 2.2 \times 10^{12}$ cm^{-1}, the values of KR range from 0.58 for Li to 1.87 for U. It will be seen from Fig. 1 that the nuclear opacity, $\sigma_a/\pi R^2$, would vary from 0.52 for Li to 0.88 for U. It will also be seen that over this range of values of KR it would be expected that σ_d will be nearly twice as large as σ_a.

If one plots the angular distribution of the diffraction scattering given by (9) (i.e., $d\sigma_d(\theta)/d\sigma_d(0)$ *versus* $kR \sin\theta$) one finds curves for the heaviest nuclei which are indistinguishable from that for an opaque nucleus (Eq. (4)), at least as far as the first minimum of the diffraction pattern. For the lighter nuclei, the form of the curve is closely the same, but with an altered scale of abscissa, corresponding to using an effective radius somewhat smaller than the true radius. The increase in the half width of the diffraction peak is zero for $KR = 1.78$ (Pb), 3.7 per cent for $KR = 1.20$ (Cu), 6.2 per cent for $KR = 0.90$ (Al) and 9.6 per cent for $KR = 0.63$ (Be). Measurements of the diffraction scattering and of the absorption are now in progress in this laboratory.

Work described in this paper was done under the auspices of the Atomic Energy Commission.

Notes

[1] R. Serber, *Phys. Rev.* **72**, 1114 (1947).

[2] L. J. Cook, E. M. McMillan, J. M. Peterson, and D. C. Sewell, *Phys. Rev.* **75**, 7 (1949).

[3] E. Amaldi, D. Bocciarelli, B. N. Cacciapuoti, and G. C. Trabacchi, *Nuovo Cimento* **3**, 203 (1946).

[4] R. Sherr, *Phys. Rev.* **68**, 240 (1945).

[5] M. L. Goldberger, *Phys. Rev.* **74**, 1268 (1948).

[6] In terms of the model being employed, the finite intercept of the R vs. $A^{\frac{1}{3}}$ line obtained from the data on the lower energy scattering could be interpreted by the more careful examination of the boundary conditions which in this case would be necessary.

[7] H. A. Bethe, *Phys. Rev.* **57**, 1125 (1940).

[8] The small dependence of K on Z/A is unimportant, as we shall see later.

7

Regularities in the Total Cross-sections for Fast Neutrons[†]

H. H. BARSCHALL

RECENTLY Miller, Adair, and others[1] have measured the total cross-sections for fast neutrons of many of the heavier elements in the energy range from about 0.1 to 3 MeV. It was found that, disregarding the effect of individual resonances, neighbouring elements show very similar variations of cross-section with energy while there are marked differences in the shape of the cross-section curves between elements of appreciably different atomic number. This behavior is shown in Fig. 1. In this figure the measured cross-sections divided by the geometrical area of the nucleus are plotted against neutron energy. The nuclear area was calculated for a nuclear radius of $1.45 \times A^{\frac{1}{3}} \times 10^{-13}$ cm. The elements are arranged according to their atomic weight A, since in the case of Te and I it was found that a smoother surface resulted from such an arrangement than if the elements were ordered according to their atomic number. No attempt has been made to include details of the fluctuations in cross-section; in particular, the behavior at thermal and epithermal energies has been ignored, since the cross-sections at the lowest energies depend primarily on the presence of individual resonances.

An interesting feature of the surface shown in Fig. 1 is the large value of the cross-section at low energies for elements around Sr. This peak appears to shift to higher energies with increasing

[†] *Phys. Rev.* **86**, 431 (1952). Work performed under the auspices of the AEC.

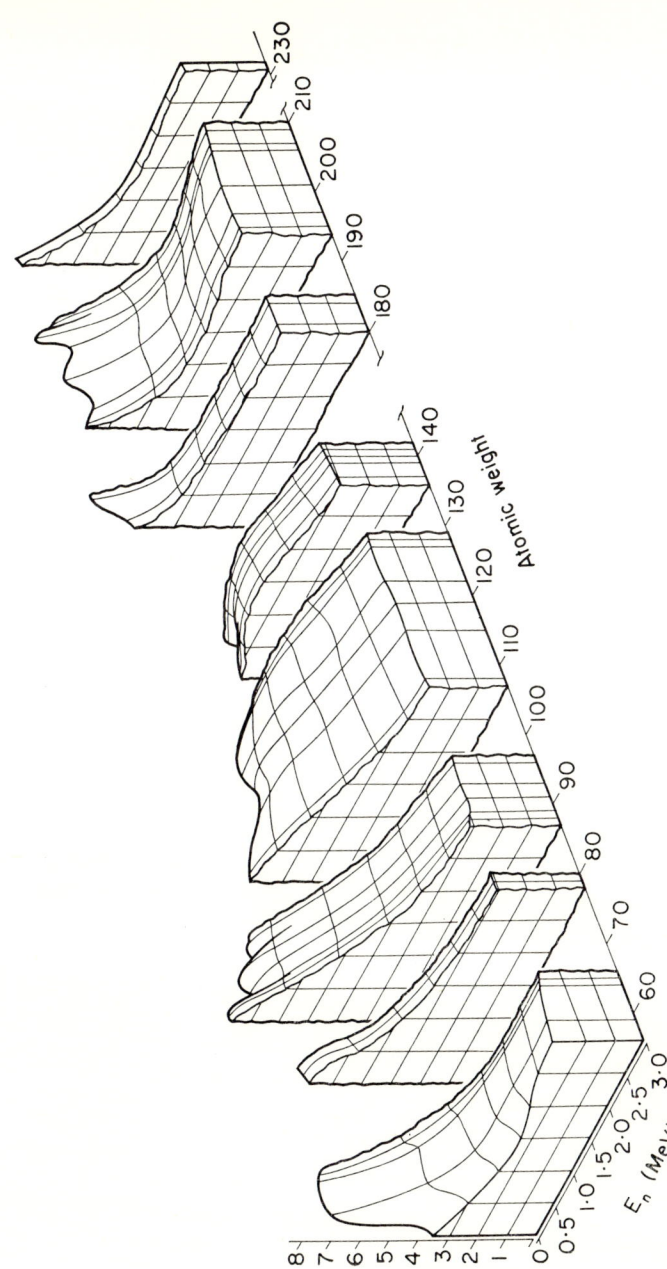

FIG. 1. Total neutron cross-sections of elements heavier than Mn as a function of neutron energy. The surface is based on measurements for the atomic weights at which straight vertical lines appear in the figure.

atomic weight. Furthermore, the cross-section of the elements heavier than Ir exhibit a minimum at neutron energies around 1 MeV.

The behavior shown in Fig. 1 is in disagreement with the continuum theory proposed by Weisskopf and his collaborators,[2] since this theory predicts a monotonic decrease of the total cross-section with energy. Following a suggestion by Wigner, Weisskopf[3] has more recently calculated the energy dependence of the cross-section on the basis of a single particle interaction and finds that variations of the total cross-section with energy similar to those shown in Fig. 1 may be obtained.

Notes

[1] Miller, Fields and Bockelman, *Phys. Rev.* **85,** 704 (1952); more complete reports will be published later.

[2] Feshbach, Peaslee, and Weisskopf, *Phys. Rev.* **71,** 145 (1947); H. Feshbach and V. F. Weisskopf, *Phys. Rev.* **76,** 1550 (1949).

[3] V. F. Weisskopf, *Bull. Am. Phys. Soc.* **27,** No. 1, 7 (1952).

8

Model for Nuclear Reactions with Neutrons†

H. FESHBACH, C. E. PORTER AND V. F. WEISSKOPF

Abstract. A simple model is proposed for the description of the scattering and the compound nucleus formation by nucleons impinging upon complex nuclei. It is shown that, by making appropriate averages over resonances, an average problem can be defined which is referred to as the "gross-structure" problem. Solution of this problem permits the calculation of the average total cross-section, the cross-section for the formation of the compound nucleus, and the part of elastic-scattering cross-section which does not involve formation of the compound nucleus. Unambiguous definitions are given for the latter cross-sections.

The model describing these properties consists in replacing the nucleus by a one-body potential which acts upon the incident nucleon. This potential $V = V_0 + iV_1$ is complex; the real part represents the average potential in the nucleus; the imaginary part causes an absorption which describes the formation of the compound nucleus. As a first approximation a potential is used whose real part V_0 is a rectangular potential well and whose imaginary part is a constant fraction of the real part $V_1 = \zeta V_0$.

This model is used to reproduce the total cross-sections for neutrons, the angular dependence of the elastic scattering, and the cross-section for the formation of the compound nucleus. It is shown that the average properties of neutron resonances, in particular the ratio of the neutron width to the level spacing, are connected with the gross-structure problem and can be predicted by this model.

The observed neutron total cross-sections can be very well reproduced in the energy region between zero and 3 MeV with a well depth of 42 MeV, a factor ζ of 0.03, and a nuclear radius of $R = 1.45 \times 10^{-13} A^{\frac{1}{3}}$ cm. The angular dependence of the scattering cross-section at 1 MeV is fairly well reproduced by the same model. The theoretical and experimental values for the ratios of neutron width to level distance at low energies and the reaction cross-sections at 1 MeV do not agree too well but they show a qualitative similarity.

† *Phys. Rev.* **96,** 448 (1954). This work was supported in part by the U.S. Office of Naval Research and the U.S. Atomic Energy Commission.

I. Introduction

This paper deals with the interaction of nuclear particles with complex nuclei in nuclear reactions. A model is proposed for the description of the energy exchange between the incoming particle and the target nucleus. The considerations are restricted to neutron reactions with incident energies between 0 and 20 MeV.

One usually describes the interaction of nuclear particles with complex nuclei by means of the concept of a compound nucleus which is formed after the nucleon has entered the nucleus. Before the striking success of the nuclear shell model was known, it was generally assumed that the quantum state formed by the particle entering the nucleus is one in which the motions of all particles are intimately coupled. We will refer to this assumption as the "strong-coupling model". These ideas led to certain general qualitative conclusions in regard to the cross-sections for nuclear reactions. Several authors[1] have attempted in previous papers to derive approximate expressions for the cross-sections of nuclear reactions with a minimum of special assumptions in addition to the main assumption of the validity of the strong-coupling model. We summarize the main results of these qualitative considerations.

1. *Particle widths.* The particle widths of nuclear resonances with respect to particle emission are related in a general way to the average spacing D of the levels of the compound nucleus. For example, the width for the emission of neutrons with zero-orbital angular momentum is given approximately by

$$\Gamma_n \approx (2/\pi)(k/K)D, \tag{1.1}$$

and the widths for the emission of other particles are equal to the above expression multiplied by the penetration factor of the potential barrier. Here k is the wave number of the incoming particle, and K is the wave number in the interior of the nucleus; K is of the order of 10^{13} cm^{-1}.

2. *Potential scattering.* The elastic scattering arises from a superposition of a resonance amplitude and a slowly varying potential scattering amplitude. The former is important only in the immed-

iate vicinity of the resonance; the latter is equal to the scattering amplitude of an impenetrable sphere of a radius approximately equal to the nuclear radius.

3. *Neutron total cross-section*. The neutron total cross-section averaged over resonances is equal to the total cross-section of a spherical potential well whose depth is such as to give rise to an internal wave number $K \sim 10^{13}$ cm^{-1} and which possesses an absorption for the incoming waves such that the waves are absorbed inside within distances of the order K^{-1}. These conditions were expressed approximately by Feshbach and Weisskopf[2] in the form of a boundary condition on the incoming wave function u/r at the nuclear boundary:

$$du/dr = -iKu.$$

Formula (1.1) and the other consequences of the strong-coupling model have been found correct as to the order of magnitude. However, as a consequence of point (3), the neutron total cross-sections when averaged over resonances should all be smooth functions of the energy which decrease monotonically with increasing energy and whose form is rather similar for all atomic numbers A. Also the dependence on A at constant energy should show a continuous, slowly increasing trend with increasing A. The measurements of neutron total cross-sections by the Wisconsin group and by others have clearly demonstrated that this is not so. The neutron total cross-sections exhibit typical deviations from the predictions of the strong-coupling model (see Fig. 3). The shape of the energy dependence of the neutron cross-sections changes significantly over the range of A; however, this change is not random but gradual. Nuclei with small differences in A show almost the same behavior. One concludes, therefore, that these characteristic shapes do not depend on detailed features of nuclear structure but on some general properties which vary slowly with A, say the nuclear radius.

The success of the shell model has cast some doubt upon the fundamental assumptions of the strong-coupling model. Does the particle necessarily form a "compound state" after entering the nucleus? The shell structure furnishes much evidence that a nucleon can move freely within the nucleus without apparently changing the quantum state of the target nucleus. This is a consequence of observations made at the ground state and at low excitation energies, and it is questionable whether this apparent absence of interaction between one nucleon and the rest is valid also at those excitation energies (~ 8 MeV) which are created in nuclear reactions with neutrons of a few MeV. Furthermore, there is some reason to believe that at higher energies, 15 MeV and up, the interaction between the entering nucleon and the target is appreciable, since the reaction cross-sections at those energies have been found[3] to be equal to the geometrical cross-sections $[\pi(R + \lambda)^2]$. Hence, for such energies it happens rarely that a neutron enters the nucleus and leaves it again without sharing its energy with the rest.

It seemed worth while, therefore, to investigate the consequences of a reduced interaction between the nucleons for the theory of nuclear reactions in the energy region of a few MeV. This reduced interaction will manifest itself in the following way: The incident nucleon can penetrate into the nucleus and move within the boundaries of the nucleus *without* forming a compound state. Hence, in this case, the target nucleus acts upon the incoming nucleon as a potential well. The actual formation of a compound state occurs only with a probability smaller than unity, once the particle has entered the nucleus. It has a finite chance of leaving the nucleus without having formed a state in which it has exchanged energy or momentum with the rest of the nucleus. The formation of the compound state then would have the aspect of an absorption. Hence, the effect of the nucleus upon the incident particle could be described as the effect of a potential well with absorption, where the absorption coefficient within the well would be an adjustable parameter. It is obvious that this description represents an oversimplification which naturally cannot reproduce all features of nuclear reactions. Specifically, it will not reproduce any resonance phenomena which are connected with the many possible quantum states of the compound system. We therefore expect that this model will at best describe only the features of nuclear reactions after averaging over the resonances of the compound nucleus.

The formulation of this attempt to construct a simple model for nuclear reactions requires a study of the definitions of the various cross-sections; in particular, the meaning of the cross-section for the formation of a compound nucleus must be clarified.

We introduce the following cross-sections: σ_t, the total cross-section, which can be split into

$$\sigma_t = \sigma_{el} + \sigma_r,$$

where σ_{el} is the elastic scattering cross-section, and σ_r is the "reaction cross-section". The former is defined as the cross-section for scattering without change of the quantum state of the nucleus. The particle leaves by the same channel by which it has entered. The elastic scattering has an angular dependence which we express by the differential cross-section $d\sigma_{el}/d\Omega$,

$$\sigma_{el} = \int \frac{d\sigma_{el}}{d\Omega}(\theta)d\Omega.$$

The reaction cross-section includes all processes in which the residual nucleus is different from or in a state different from that of the target nucleus. These are all processes whose exit channels differ from the entrance channel. It will be practical later on to subdivide the elastic cross-section into two parts:

$$\sigma_{el} = \sigma_{se} + \sigma_{ce}.$$

We call the second part, σ_{ce}, the "compound elastic" cross-section. It is the part of the elastic scattering which comes from the formation of the compound nucleus and the subsequent emission of the incident particle into the entrance channel. The first part we call "shape elastic" cross-section; this is the part of the elastic scattering which occurs without the formation of a compound. The exact definition of this split will be given in section II. We note that such definitions will be possible only for the average cross-sections, averaged over an energy interval containing many resonances, if such resonances are present.

On the basis of the compound nucleus assumption, we consider all actual reactions to occur after compound formation. Hence, we introduce a cross-section σ_c of compound nucleus formation:

$$\sigma_c = \sigma_{ce} + \sigma_r,$$

and obtain, naturally,

$$\sigma_t = \sigma_{se} + \sigma_c.$$

The nuclear model which we propose here is expected to predict only the cross-sections σ_{se} and σ_c. It considers only the conditions in the entrance channel, that is, in that part of the phase space in which the target nucleus is in its initial state. Hence, the compound nucleus formation is considered as an *absorption* of the incident beam, although part of it, namely σ_{ce}, leads to an elastic scattering process. The model consists in describing these conditions by means of a one-particle problem. The nucleus is replaced by a complex potential,

$$V = V_0 + iV_1, \qquad (1.2)$$

acting upon the incoming neutron. The scattering which the neutron suffers in (1.2) should reproduce the shape elastic scattering σ_{se}; and the absorption which is caused by the imaginary part V_1 should reproduce the compound nucleus formation.

It is probable that the potential functions in (1.2) vary somewhat with the incident energy. For example, one might expect an increase of the imaginary part with increasing energy. If an approximate description of the facts is possible by means of a potential (1.2), the shape of the potential will be indicative of the type of nuclear interaction which a neutron suffers in the nucleus. The real part V_0 would describe the average potential energy of the neutron within the nucleus, and its shape would give indications as to the form of the potential "well" inside the nucleus. It is similar to the potential encountered in the shell model of the nucleus, although we do not pretend that an incident neutron of several MeV is faced with exactly the same potential which acts upon the nucleus in the ground state. The imaginary part V_1 would indicate the strength and location of the processes that lead to an energy exchange between the incoming neutron and the target nucleus.

We expect the potential V to depend in a simple way upon the mass number A. Its dependence on r should be similar for all nuclei. The simplest choice would be a square-well potential:

$$V_0 = -U \quad \text{for} \quad r < R,$$
$$V_0 = 0 \quad \text{for} \quad r > R,$$
$$V_1 = \zeta V_0.$$

In general, we might express it in the form $V = V(r/R)$, $R = r_0 A^{\frac{1}{3}}$. However, there might be a region near $r = R$ in which the features depend on r and not on (r/R); the thickness of that part of the potential which represents the surface might be independent of the radius.

With a given $V(r)$ and its dependence on A, it is possible to calculate the cross-sections σ_t, σ_{sc}, and σ_c, each as functions of energy and mass number, and also the angular dependence of the scattering. The next section contains the definitions of the cross-sections involved, and the following sections describe the technique of calculating the cross-sections and their comparison with experimental material.

II. Theory of Average Cross-sections

All nuclear cross-sections exhibit strong fluctuations with energy which are generally referred to as resonances, especially in the lower part of our energy range. As the energy increases, the width of the resonances increases too; and, for not too light nuclei, the width becomes comparable or larger than the level distance at energies above a few MeV. Hence, we find the cross-sections at higher energies to be smooth functions of energy with little fluctuation. We will refer to the lower-energy region as the "resonance region" and the upper as the "continuum region".

The behaviour of the cross-sections in the resonance region does not lend itself to a description by a simple one-particle potential (1.2) because of the rapid fluctuations with energy. However, the averages of the cross-sections taken over an interval I, which includes many resonances, will be shown to be the cross-sections belonging to a new scattering problem with slowly varying phases, which we will call the "gross-structure" problem. In this problem it is possible to define cross-sections for the formation of a compound nucleus which also includes the compound elastic scattering. *It is this gross-structure problem and not the actual rapidly varying cross-sections which we intend to describe by means of a one-particle problem with the potential* (1.2).

We bombard a nucleus X with particles a and consider the total cross-section σ_t, the elastic cross-section σ_{el}, and the reaction cross-section σ_r. $\sigma_t = \sigma_{el} + \sigma_r$. Each of these cross-sections will be subdivided into their parts coming from different angular momenta l, e.g.,

$$\sigma_t = \sum_l \sigma_t^{(l)}. \tag{2.1}$$

These cross-sections can be expressed in terms of the amplitudes of the wave which describes the situation in the entrance channel. We consider the subwave u_l/r in the entrance channel with the orbital angular momentum l (r is the channel coordinate), and we write the wave in the form for

$r \to \infty$,

$$\varphi_l \to \text{const}[\exp(-i(kr - \tfrac{1}{2}l\pi)) - \eta_l \exp(+i(kr - \tfrac{1}{2}l\pi))]. \tag{2.2}$$

The complex reflection factor η_l is connected with the complex phase shift φ_l by $\eta_l = \exp(2i\varphi_l)$, and the cross-sections are given by the well-known expressions for the elastic cross-section:

$$\sigma_{el}^{(l)} = \pi \lambda^2 (2l + 1) \mid 1 - \eta_l \mid^2, \tag{2.3}$$

and for the reaction cross-section,

$$\sigma_r^{(l)} = \pi \lambda^2 (2l + 1)(1 - \mid \eta_l \mid^2), \tag{2.3a}$$

where $\lambdabar$ is the wavelength of the incoming particle divided by 2π.

The reflection factor η_l is a complicated function of the energy of the incoming particle. It exhibits rapid fluctuations coming from the numerous close-spaced resonances of the compound nucleus. We will make the assumption that one can average over these fluctuations; that is, we assume that the average reflection factor,

$$\bar{\eta}_l(\epsilon) = \frac{1}{I} \int_{\epsilon-I/2}^{\epsilon+I/2} \eta_l(\epsilon') d\epsilon', \qquad (2.4)$$

is a smooth function of ϵ if the interval I contains many close-spaced resonances. We also define average cross-sections in the same way, and we can write

$$\bar{\sigma}_{el}^{(l)} = \pi\lambdabar^2(2l+1)\, \overline{\mid 1 - \eta_l \mid^2},$$
$$\bar{\sigma}_r^{(l)} = \pi\lambdabar^2(2l+1)\, (1 - \overline{\mid \eta_l \mid^2}), \qquad (2.5)$$

where the bar over an expression signifies its average over the interval I. It is also assumed that I is much smaller than the energy ϵ such that slowly varying functions of ϵ, like $\lambdabar^2$, need not be averaged.

One can easily verify the following relations:

$$\bar{\sigma}_{el}^{(l)} = \pi\lambdabar^2(2l+1)\{\mid 1 - \bar{\eta}_l \mid^2 - \mid \bar{\eta}_l \mid^2 + \overline{\mid \eta_l \mid^2}\}, \qquad (2.6)$$

and especially

$$\bar{\sigma}_t^{(l)} = \pi\lambdabar^2(2l+1)\{\mid 1 - \bar{\eta}_l \mid^2 + 1 - \mid \bar{\eta}_l \mid^2\}. \qquad (2.7)$$

Hence, the average total cross-section depends only upon the average reflection factor (2.4). [This follows directly from the fact that the total cross-section is a *linear* function of the real part of the phase η_l.]

We now divide the average elastic cross-section into two parts, the "shape elastic" cross-section $\sigma_{se}^{(l)}$ and the "compound elastic" cross-section[4] $\sigma_{ce}^{(l)}$, by writing

$$\sigma_{se}^{(l)} = \pi\lambdabar^2(2l+1)\mid 1 - \bar{\eta}_l \mid^2,$$
$$\sigma_{ce}^{(l)} = \pi\lambdabar^2(2l+1)\{\overline{\mid \eta_l \mid^2} - \mid \bar{\eta}_l \mid^2\}. \qquad (2.8)$$

Furthermore, we combine $\sigma_{ce}^{(l)}$ and $\bar{\sigma}_r^{(l)}$ into a new cross-section $\sigma_c^{(l)}$, which we call the cross-section for the formation of the compound nucleus

$$\sigma_c^{(l)} = \sigma_{ce}^{(l)} + \bar{\sigma}_r^{(l)} = \pi\lambdabar^2(2l+1)\{1 - |\bar{\eta}_l|^2\}. \quad (2.9)$$

We can see from (2.3) and (2.3a) that $\sigma_{se}^{(l)}$ and $\sigma_c^{(l)}$ have just the form of a scattering and a reaction-cross section of a new and different problem, whose phase is the slowly varying function $\bar{\eta}_l$. In other words, by replacing η_l with $\bar{\eta}_l$, we obtain a new problem, which we have called the "gross-structure problem". The elastic scattering cross-section σ_{se} of *this* problem is only part of the actual scattering; it is the "shape elastic" scattering. The other part, the "compound elastic", appears incorporated into the absorption or reaction cross-section σ_c of the gross-structure problem together with the actual cross-section.

One is therefore led to consider the "compound elastic" scattering as that part which comes from the formation of the compound nucleus and its subsequent decay into the entrance channel, hence its incorporation into σ_c. After the averaging, σ_{ce} appears as part of the absorption from the incoming beam, which corresponds to the idea that the formation of the compound nucleus can be considered as an absorption whatever happens afterwards, re-emission or not.

It is the gross structure problem which we intend to reproduce by the interaction of the incident particle with the potential (1.2). The resulting scattering cross-section should represent the shape elastic scattering, and the resulting absorption cross-section should represent the compound formation. The latter contains the part σ_{ce} of the actual scattering.

When the energy is high enough above the resonance region that the continuum region is reached, the cross-sections and phases are no longer rapidly varying functions of energy. Then the gross-structure problem is equal to the actual one and $\bar{\eta}_l = \eta_l$. It follows from (2.9) that $\sigma_{ce} = 0$. One also can see this from an application of the compound nucleus assumption to the continuum region. The overlap of the resonances can be interpreted as a consequence

MODEL FOR NUCLEAR REACTIONS WITH NEUTRONS 237

of the fact that the probability $\Gamma_\alpha{}^s$ of the decay of the compound nucleus in the state s into the entrance channel α is much smaller than the probability of the decay into other channels. This follows from the well-known relation that any channel width $\Gamma_\alpha{}^s$ cannot be larger than $D/2\pi$ (D is the distance between resonances of the same J value). Hence, if the total width is much larger than D, the contribution to Γ from decays other than the one through α must be overwhelming. In the continuum region, therefore, the cross-section for the formation of the compound nucleus is identical to the average reaction cross-section $\bar\sigma_r$, and σ_{ce} is negligible.

We now illustrate the averaging process described above by using cross-sections as given by the Breit–Wigner formula. We consider a nucleus with resonances at the energies ϵ_s, and we restrict our considerations to neutrons with $l = 0$. We also restrict the discussion to low energies so that the following two magnitudes are small: One is kR and the other is Γ/D, with R the nuclear radius, and Γ and D the average values of the total width of and the distance between neutron resonances.

We have derived in Appendix 1 exact and approximate expressions for the scattering amplitude η_0 and for the cross-sections in this energy region. For the present purposes, we will use the following form (see A.14b), which is valid in a region D_s including a resonance ϵ_s as indicated:

$$\eta_0 = e^{-2ikR'(\epsilon)} \left(1 - \frac{i\Gamma_\alpha{}^s}{\epsilon - \epsilon_s + i\Gamma^s/2}\right) + \eta_0{}^*,$$

$$\eta_0{}^* = e^{-2ikR'(\epsilon)} \left(\frac{\Gamma_\alpha{}^s}{\epsilon - \epsilon_s + i\Gamma^s/2} G_1 + iG_2 + G_3\right), \quad (2.10)$$

for

$$\epsilon_s + \epsilon_{-1} < 2_s\epsilon < \epsilon_{s+1} + \epsilon_s.$$

Here R' is a length and a slowly varying function of the energy. (A function is slowly varying if it changes value appreciably only over intervals large compared to D.) The length R' is of the order of magnitude of nuclear dimensions. It plays the role of a scattering length and takes on both positive and negative values. The quantities Γ_α^s and Γ^s are the partial width and the total width, respectively. The terms G_1, G_2, and G_3 are real functions of ϵ of the following order of magnitude:

$$G_1 \sim \Gamma/D, \; G_2 \sim (\Gamma_\alpha/D)[(\Gamma/D) + kR], \; G_3 \sim \Gamma_\alpha/D, \quad (2.11)$$

where the omission of the superscript signifies the average value of the magnitude in the interval I.

The first term in η_0 incorporates the contribution from the resonance level ϵ_s, whereas η_0^* contains the contribution from the other resonances; the first term in η_0^* represents interference effects between the resonance ϵ_s and other resonances. It will appear later that η_0^* contributes negligibly to the average of η_0.

The cross-sections in the immediate neighbourhood of the resonance ($|\epsilon - \epsilon_s| \ll D_s$) follow from (2.5) and (2.10) by neglecting η_0^*, since, in that region, they contribute terms much smaller than the others.

$$\sigma_r^{(0)} = \pi\lambda^2 \frac{\Gamma_\alpha^s(\Gamma^s - \Gamma_\alpha^s)}{(\epsilon - \epsilon_s)^2 + (\Gamma^s/2)^2},$$

$$\sigma_{el}^{(0)} = \pi\lambda^2 \left| (e^{2ikR'} - 1) + \frac{i\Gamma_\alpha^s}{\epsilon - \epsilon_s + i\Gamma^s/2} \right|^2, \quad (2.12)$$

$$|\epsilon - \epsilon_s| \ll D.$$

The reaction cross-section is just the sum over β of the one-level Breit–Wigner cross-sections

$$\sigma_{\alpha\beta}^{(0)} = \frac{\Gamma_\alpha^s \Gamma_\beta^s}{(\epsilon - \epsilon_s)^2 + (\Gamma^s/2)^2} \pi\lambda^2 \quad (2.13)$$

for the reaction leading from the entrance channel α to an exit channel β. (Γ_β^s is the partial width of decay into the channel β.)

The elastic cross-section contains a "potential" scattering amplitude

$$P = e^{2ikR'(\epsilon)} - 1,$$

which corresponds to a scattering at a hard sphere of a radius R', where R' is not identical to but only of the order of magnitude[5] of the nuclear radius and is a slowly varying function of the energy.

We now determine the average value of the scattering amplitude η_0 over the resonances in the interval I:

$$\bar{\eta}_0 = \frac{1}{I} \int_I \eta_0 d\epsilon = \left\langle \frac{1}{D_s} \int_{D_s} \eta_0(\epsilon) d\epsilon \right\rangle_I,$$

where the symbol $\langle \ \rangle_I$ signifies an average taken over all resonances within I. The random position of resonances allows us to write

$$\bar{\eta}_0 = \frac{1}{D} \int_{\epsilon_s - D/2}^{\epsilon_s + D/2} \eta_0(\epsilon) d\epsilon, \tag{2.14}$$

where D is the average level distance within the interval I; typical average values of Γ^s and Γ_α^s should be used in the expression (2.10) for η_0.

Evaluation of (2.14) gives

$$\bar{\eta}_0 = e^{-2ikR'}[1 - (\pi \Gamma_\alpha / D)], \tag{2.15}$$

when all magnitudes of the order $(\Gamma_\alpha \Gamma)/D^2$ or $(\Gamma_\alpha/D)kR'$ or smaller are neglected. It is seen in each interval D_s that the main contribution to the average comes from the main resonance. The contribution of neighbouring resonances which are expressed by η_0^* in (2.10) contribute only to expressions which are smaller than (2.14) by a factor of the order Γ/D or kR.

We now use (2.14) for the calculation of the "shape elastic" scattering and get, with the help of (2.8),

$$\sigma_{se}^{(0)} = \pi \lambda^2 \, | \, (e^{2ikR'} - 1) + \pi \Gamma_\alpha / D \, |^2. \tag{2.16}$$

For small kR', this becomes

$$\sigma_{se}^{(0)} = 4\pi R'^2 [1 + (\pi \Gamma_\alpha / 2kR'D)^2]. \tag{2.17}$$

The magnitude $[\pi \Gamma_\alpha / 2(kR'D)]^2$ is usually rather small. [It is of the order of 10^{-2}; see, for example, the estimate in Blatt and Weisskopf,[5] chap. VIII, eq. (7.14).] Hence, $\sigma_{se}^{(0)}$ is very nearly equal to $4\pi R'^2$ for $kR' \ll 1$.

240 NUCLEAR REACTIONS

We get the cross-section for the formation of the compound nucleus according to (2.9),

$$\sigma_c^{(0)} = 2\pi^2 \lambdabar^2 (\Gamma_\alpha/D) (1 - \tfrac{1}{2}\pi\Gamma_\alpha/D). \tag{2.18}$$

The average total cross-section then becomes

$$\sigma_t^{(0)} = \sigma_{se}^{(0)} + \sigma_c^{(0)} = 4\pi R'^2 + 2\pi^2 \lambdabar^2 \Gamma_\alpha/D. \tag{2.19}$$

The second term in this expression is proportional to $(1/v)$.

It is interesting to compare σ_c with the average of σ_r, which, according to (2.12), is

$$\bar\sigma_r^{(0)} = 2\pi^2 \lambdabar^2 (\Gamma_\alpha/D)(\Gamma - \Gamma_\alpha)/\Gamma. \tag{2.20}$$

Hence, the difference between the two, the "compound elastic" scattering is, [neglecting the small factor $\pi\Gamma_\alpha/(2D)$]

$$\sigma_{ce}^{(l)} = 2\pi^2 \lambdabar^2 \Gamma_n^2/D\Gamma. \tag{2.21}$$

This is just the average of that part of the elastic scattering (2.12) which corresponds to the resonance amplitude only, namely, of

$$\sigma = \pi\lambdabar^2 \frac{(\Gamma_\alpha^s)^2}{(\epsilon - \epsilon_s)^2 + (\Gamma^s/2)^2}.$$

It is the cross-section which one would get for the re-emission into the entrance channel from the Breit–Wigner expression (2.13).

It is significant that expressions (2.15), (2.18), and (2.19) do not contain the total width Γ but only the channel width Γ_α. The "gross" properties (total, shape elastic, and compound nucleus formation cross-sections) are independent of the nature of the other exit channels. They would remain unchanged, for example, if the exit channels $\beta \neq \alpha$ were closed. It would only increase σ_{ce} at the expense of $\bar\sigma_r$, as seen in (2.20) and (2.21). This is connected with the fact that a change of Γ with constant Γ_α changes only the width of the resonance, but not its area.

At the energies considered here, the cross-section for the formation of the compound nucleus contains only magnitudes (Γ_α and D), which can be determined by studying the neutron resonances. Hence, investigations of slow neutron resonances are useful to check the theoretical predictions of σ_c at low energy. The "shape-elastic" scattering, on the other hand, in this energy region is almost entirely given by $4\pi R'^2$ and is therefore essentially independent of the neutron resonance values. Apart from the small correction $\pi^3 \lambda^2 (\Gamma_a/D)^2$, it is equal to the potential scattering as shown in (2.12) and, therefore, can be measured also by studying the cross-sections near and between resonances.

III. Potential-well Model

In this section we shall employ a potential-well model to determine the gross-structure cross-sections. We have adopted for the purposes of a preliminary survey the simplest type of potential well:

$$V = -V_0(1 + i\zeta), \quad r < R,$$
$$V = 0, \qquad\qquad r > R, \qquad (3.1)$$

where V_0 and ζ are constants and R is the nuclear radius. The use of the complex potential is necessary to obtain nonzero values for the cross-section for the formation of the compound nucleus. A similar model in which $\zeta \sim 1$ was employed by Bethe.[6] Fernbach, Serber, and Taylor[7] have used the same model in order to describe nuclear scattering at very high energies. A model in which $\zeta = 0$ was used by Ford and Bohm[8] in discussing zero-energy cross-sections. It is essential that the crudeness of this model be emphasized. We have, for example, omitted any spin–orbit terms which play an important role in the shell model, but which we expect will not affect the over-all qualitative features which we seek here. The constants in (3.1) may well turn out to be energy dependent. We particularly expect this for ζ, since we know that $\bar{\sigma}_r$ is large at high energies, while the success of the shell model indicates that ζ should be zero for the ground states of nuclei.

We give some of the details of the calculations with potential (3.1). For each l we calculate the value of the logarithmic derivative,

$$f_l = R(u_l'/u_l)_{r=R}. \qquad (3.2)$$

The average reflection factor $\bar{\eta}_l$ is then

$$\bar{\eta}_l = e^{-2i\delta_l}\left(1 - \frac{2s_l}{M_l + iN_l}\right), \tag{3.3}$$

where

$$\delta_l = \tan^{-1}(-j_l(x)/n_l(x)), \tag{3.4a}$$

$$\Delta_l + is_l = 1 + xh_l'(x)/h_l(x), \tag{3.4b}$$

$$M_l = s_l - \mathrm{Im}\, f_l, \quad N_l = -\Delta_l + \mathrm{Re}\, f_l. \tag{3.4c}$$

The functions j_l, n_l, and h_l are the spherical Bessel, Neumann, and Hankel functions, respectively, while x is, as usual, kR.[9] $h_l'(x)$ is the derivative of $h_l(x)$ with respect to x; Δ_l and s_l are both real magnitudes and are defined as the real and imaginary part of the expression on the right of (3.4b).

For potential (3.1), f_l may be written down directly

$$f_l = 1 + Xj_l'(X)/j_l(X), \tag{3.5}$$

where

$$X^2 = x^2 + X_0^2(1 + i\zeta), \quad X_0^2 = (2m/\hbar^2)V_0 R^2$$

This is, however, not the most convenient form for determining the real (Re) and imaginary (Im) parts of f_l. We have instead employed recurrence relations for these quantities based on recurrence relations for j_l.

For $l = 0$, we get

$$f_0 = X \cot X,$$

$$\mathrm{Re}\, f_0 = \frac{X_1 \sin 2X_1 + X_2 \sinh 2X_2}{\cosh 2X_2 - \cos 2X_1}, \tag{3.6}$$

$$\mathrm{Im}\, f_0 = \frac{X_2 \sin 2X_1 - X_1 \sinh 2X_2}{\cosh 2X_2 - \cos 2X_1},$$

where $X = X_1 + iX_2$. The recurrence relations which follow from

$$f_l = \frac{X^2}{l - f_{l-1}} - l$$

are
$$\operatorname{Re} f = \frac{(X_1^2 - X_2^2)(l - \operatorname{Re} f_{l-1}) - 2X_1 X_2 \operatorname{Im} f_{l-1}}{(l - \operatorname{Re} f_{l-1})^2 + (\operatorname{Im} f_{l-1})^2} - l, \quad (3.7)$$

$$\operatorname{Im} f_l = \frac{(X_1^2 - X_2^2) \operatorname{Im} f_{l-1} + 2X_1 X_2(l - \operatorname{Re} f_{l-1})}{(l - \operatorname{Re} f_{l-1})^2 + (\operatorname{Im} f_{l-1})^2}. \quad (3.8)$$

The asymptotic expression for f_l,

$$f_l \xrightarrow[X \to \infty]{} X \cot(X - \tfrac{1}{2} l \pi), \quad (3.9)$$

unfortunately cannot be generally employed. The fractional error in (3.9) is $l(l+1)/(X \sin 2X)$, from which we learn that (3.9) is not sufficiently accurate for $l \geqslant 2$, while for $l = 1$ it will fail for small X or for $X = n\pi$.

The total cross-section, as well as the cross-section for the formation of the compound nucleus, may be easily obtained

$$\frac{\bar{\sigma}_t^{(l)}}{\pi R^2} = \frac{4}{x^2}(2l+1)\left[\sin^2 \delta_l + s_l \frac{M_l \cos 2\delta_l - N_l \sin 2\delta_l}{M_l^2 + N_l^2}\right],$$

$$\frac{\bar{\sigma}_c^{(l)}}{\pi R^2} = \frac{4}{x^2}(2l+1)s_l\left[\frac{-\operatorname{Im} f_l}{M_l^2 + N_l^2}\right] \equiv \frac{(2l+1)T_l}{x^2}, \quad (3.10)$$

$$\bar{\sigma}_t = \Sigma_l \bar{\sigma}_t^{(l)}, \quad \sigma_c = \Sigma_l \sigma_c^{(l)},$$

where the T_l may be interpreted as penetrabilities.

These cross-sections will have characteristic large-scale resonances, which are present in the experimental data. In the $l = 0$ case, these resonances occur when

$$X_1 = (X_0^2 + x^2)^{\frac{1}{2}} = (n + \tfrac{1}{2})\pi + \frac{X_0^2 \zeta^2}{2(2n+1)\pi},$$

where n is an integer and where we have assumed that $\zeta X_0^2/n\pi \ll 1$. The width of the large-scale resonance is $2x\hbar^2/mR^2$, which in the experimental range is of the order of MeV. For a given energy, the $l = 0$ cross-section will give maxima as a function of R. The width of these maxima against changes in R is approximately $(2xR/X^2)$, independent of R.

The angular distribution for shape elastic scattering is

$$\frac{d\sigma_{se}}{d\Omega} = \frac{\lambda^2}{4} \mid \sum_l (2l+1)(1-\bar{\eta}_l)P_l(\cos\theta) \mid^2.$$

Therefore

$$\frac{1}{R^2} = \frac{d\sigma_{se}}{d\Omega} = (\text{Re } \Sigma)^2 + (\text{Im } \Sigma)^2, \tag{3.11}$$

where

$$\text{Im } \Sigma = \frac{x}{4} \sum_l \frac{\sigma_t^{(l)}}{\pi R^2} P_l(\cos\theta), \tag{3.12}$$

$$\text{Re } \Sigma = \frac{1}{2x} \sum_l (2l+1) \left[\sin 2\delta_l - 2s_l \frac{M_l \sin 2\delta_l + N_l \cos 2\delta_l}{M_l^2 + N_l^2} \right].$$

Before we can compare the theory with the experimental data on angular distributions, it is necessary to add the compound elastic scattering. From our general qualitative ideas, we may break the process up into the formation of the compound nucleus and the re-emission of the incident particle into a particular l state, which will naturally have a very definite associated angular distribution. The result is particularly simple in the case of a target nucleus of spin zero and an incident particle of spin zero, since here the angular momentum of the incident particle cannot change in an elastic scattering process. We may therefore write

$$\frac{d\sigma_{ce}}{d\Omega} = \sum \sigma_c^{(l)} \mid Y_{l0} \mid^2 w_l, \tag{3.13}$$

where Y_{l0} are the normalized spherical harmonics and w_l is the probability that the compound nucleus formed by the absorption of a particle of angular momentum l will decay by emission of the same particle without loss of energy or change in angular momentum.

This simple result cannot be applied to the neutron case because of the possibility of spin changes of the neutron and re-orientation of the spin of the target nucleus without any change in the energy of either the neutron or the target nucleus. The formalism which needs to be used here has been worked out by Hauser and Feshbach[10] and by Wolfenstein.[11] The target nucleus and neutron system is now characterized by the spin of the target nucleus I, its z component m, and spin of the neutron i, the channel spin s ($\mathbf{s} = \mathbf{i} + \mathbf{I}$), the angular momentum of the incident neutron l, and its z component which is zero, and of course the parity of the system. The compound nucleus will have a total angular momentum J, z component m, and will decay into a residual nucleus of spin I' and a particle of spin i'. These form a final channel spin s' ($\mathbf{s}' = \mathbf{i}' + \mathbf{I}'$), z component $m - m'$. The system will have an angular momentum l', z component m'.

To obtain the desired cross-section, we must now introduce the assumptions of the statistical nuclear theory. We assume that, upon appropriate averaging, the various J levels do not interfere, that there is no residual interference between the various l's which can form the given compound state J, or between the various l's into which it can decay. We then break up the process of compound elastic scattering into the cross-section for the formation of the compound nucleus in state J, with incident particles of angular momentum l, multiplied by the probability that it will decay by emission of a particle of angular momentum l', leaving the residual nucleus in the ground state with spin I. The formation process is given first by the cross-section for the formation of the compound nucleus which, because of our simple assumption (3.1), depends only on l and is $\sigma_c^{(l)}$. This must be multiplied by the probability of forming the system with angular momentum J, using incident particles of angular momentum l. Again, because of the absence of any spin-dependent forces, this is simply the square of the Clebsch–Gordan coefficient $|(ls0m|lsJm)|^2$. On the emission side, we will need the probability of forming J with particles of angular momentum l'. This is given by $|(l's'm'm - m'|l's'Jm)|^2$. We need the relative probability of decay with emission of l' particles leaving the nucleus in the ground state which we will denote by $w(l') \leqslant 1$. The limitations of the relative probability of different kinds of emission arising from angular momentum conservation are contained in the Clebsch–Gordan coefficients. The function w contains all the other dependence. Because of our assumption of spin-independent forces in eq. (3.1), it will depend only on l' and the parity of the system. The angular distribution of the emitted particles is $|Y_{l',m'}|^2$. Combining these results, we have

$$\frac{d\sigma_{ce}}{d\Omega} = \frac{1}{(2i+1)(2I+1)} \Sigma \, \sigma_c^{(l)} \, |(ls0m \, | \, lsJm)|^2$$
$$\times |(l's'm'm - m' \, | \, l's'Jm)|^2 w(l') \, | \, Y_{l'm'}|^2. \quad (3.14)$$

The indicated sums are over m, m', s, s', l, l', and J. The spin factor in front arises from the average over initial spin states, which involves the sum over m and m'.

By employing methods due to Racah[12] and discussed by Blatt and Biedenharn,[13] the sums over m and m' may be performed yielding

$$\frac{d\sigma_{ce}}{d\Omega} = \frac{1}{4\pi(2I+1)(2i+1)} \sum \frac{\sigma_c^{(l)} w(l')}{2l+1}$$
$$\times Z(lJlJ; sL)Z(l'Jl'J; s'L)P_L(\cos\theta), \quad (3.15)$$

where the Z factors are defined by Biedenharn, Blatt, and Rose[14] and for which tables [15] are available. The sums are over J, l, l', and L. Only even L will occur. This result is given in footnote 10.[16] We have not introduced a specific notation to describe the role of parity, so it should be understood that parity is conserved both in the formation and in the decay of the compound nucleus.

The total compound elastic cross-section may be easily evaluated from (3.15) and gives the expected result

$$\sigma_{ce} = \sum \frac{2J+1}{(2I+1)(2i+1)(2l+1)} \sigma_c^{(l)} w(l').$$

Expression (3.15) simplifies considerably in two special cases (a) $I \gg 1$ and (b) $I = 0$. In case (a), it follows from the sum rule (see reference 10),

$$\Sigma_{s'l'}(2s'+1) \, | \, Z(lJlJ; sL)Z(l'Jl'J; s'L) \, | \, P_L = (2J+1)^2(2l+1), \quad (3.16)$$

that $d\sigma_{ce}/d\Omega$ is approximately independent of angle. We note that if $I \gg 1$, the factor $2s' + 1$ is approximately a constant, the error being of the order of $(1/I)$. In case (b) we note that $s = s' = \frac{1}{2}$ and that $l = l'$ because of parity conservation. We therefore find for this case (placing $i = \frac{1}{2}$)

$$\frac{d\sigma_{ce}}{d\Omega} = \sum_{l,L}' \frac{\sigma_c^{(l)} w(l)}{4\pi(4l+2)} [Z^2(l, l+\tfrac{1}{2}, l, l+\tfrac{1}{2}; \tfrac{1}{2}, L)$$
$$+ Z^2(l, l-\tfrac{1}{2}, l, l-\tfrac{1}{2}; \tfrac{1}{2}, L)] P_L. \tag{3.17}$$

The factors w which may be computed as outlined in note 10 depend on the details of the levels of the residual nucleus. There it is shown that

$$w(l') = T_{l'}(E)/\Sigma_{pqs} T_p(E_q'). \tag{3.18}$$

The quantity $w(l')$ lies between 0 and 1. The values of T_p are calculated in reference 10 under the assumption of strong coupling. The ideas underlying the present theory would change these factors to those given by eq. (3.10). Since the compound elastic scattering is not very large compared to the shape elastic, we have only determined the upper limit for σ_{ce}, which is given by putting $w(l') = 1$. We expect σ_{ce} to be near this upper limit at energies for which there is little inelastic scattering or capture, and to be near zero when inelastic scattering or other nuclear reactions are appreciable.

IV. Isolated Resonances

We should like to establish a correspondence between the parameters describing a single compound nucleus resonance and the parameters which describe the average potential (3.1). This is most easily done for the low-energy case. We evaluate the cross-sections for very low energy on the basis of the potential (3.1) and by comparing them with the expressions for the average cross-sections which were derived in section II in terms of the resonance parameters. The only two resonance parameters entering here are the ratio Γ_α/D of neutron width[17] to level distance and the radius R' of the potential scattering.

We start with the evaluation of the results from (3.1). The only contribution comes from $l = 0$ and we get from (3.3)

$$\bar{\eta}_0 = e^{-2ix}\frac{f_0 + ix}{f_0 - ix} = e^{-2ix(1-\alpha)}, \qquad (4.1)$$

where α is a complex number:

$$\alpha = (1/x) \tan^{-1}(x/f_0) \cong 1/f_0,$$

and f_0 is given by the expressions (3.6).

This should be compared with (2.14) in order to express the two relevant magnitudes R' and Γ_α/D in terms of X_1 and X_2. Equating (2.14) and (4.1) gives in the limit of $k \to 0$, a limit which also implies $\Gamma_\alpha/D \to 0$:

$$R' = R(1 - \alpha_1), \quad (\pi/2x)\,(\Gamma_\alpha/D) = \alpha_2,$$

where α_1 and α_2 are the real and imaginary parts of α.

From (3.6) we can easily obtain the following relations:

$$\alpha = f_0^{-1} = \alpha_1 + i\alpha_2,$$

$$\alpha_1 = \frac{1}{|X|^2} \frac{X_2 B - X_1 A \sin 2X_1}{B^2 + 2A \cos^2 X_1},$$

$$\alpha_2 = \frac{1}{|X|^2} \frac{X_1 B - X_2 A \sin 2X_1}{B^2 + 2A \cos^2 X_1},$$

with $A = 1/(2 \cosh^2 X_2)$, $B = \tanh X_2$.

We now distinguish two limiting cases: the cases of strong and weak coupling. In the first case the absorption is so strong that the neutron is completely absorbed in a distance of a nuclear radius within nuclear matter: $\exp(-X_2) \ll 1$. In the case of weak coupling we assume $X_2 \ll 1$.

Hence we get, for strong coupling: $A \to 0$, $B \to 1$, and

$$\alpha_1 = X_2/|X|^2 = 1/X'_2, \quad \alpha_2 = X_1/|X|^2 = 1/X'_1,$$

and $R' = R(1 - 1/X'_2)$, $\Gamma_\alpha/D = 2x/\pi X'_1$. The length R' is almost equal to R since $(X'_2)^{-1}$ is a small magnitude. The expression for Γ_α/D is the same as that used by Feshbach, Peaslee, and Weisskopf

MODEL FOR NUCLEAR REACTIONS WITH NEUTRONS 249

with the only exception that X_1' replaces X_1. The former magnitude is somewhat larger than X_1.[18]

Strong coupling therefore leads essentially to the same results as Feshbach, Peaslee, and Weisskopf: The potential scattering length is roughly equal to R and $\Gamma_\alpha/D = 2(x/\pi X_1')$.

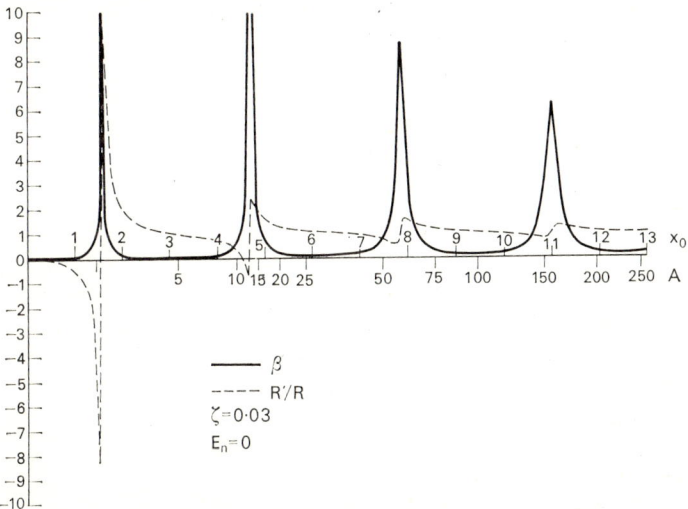

FIG. 1. Potential scattering length R' and the ratio Γ_α/D of the neutron width to the level distance at low energy as a function of $X_0 = K_0 R$ for $\zeta = 0.03$. R' is plotted in units of R and Γ_α/D is given in the form of the parameter $\beta = (\pi/2)(V_0/\epsilon)^{\frac{1}{2}} (\Gamma_\alpha/D)$, where ϵ is the energy of the neutron. The atomic weight scale corresponding to X_0 is shown also for a potential-well depth $V_0 = 42$ MeV and radii $R = 1.45 \times 10^{-13} \, A^{\frac{1}{3}}$ cm.

In the weak coupling approximation we get

$$A = \tfrac{1}{2}, \quad B = X_2 \ll 1, \quad |X|^2 = X_1^2,$$

and hence

$$R' = R\left(1 - \frac{1}{2X_1} \frac{\sin 2X_1}{X_2^2 + \cos^2 X_1}\right),$$

$$\frac{\Gamma_\alpha}{D} = \frac{2x}{\pi X_1} \beta, \quad \beta = X_2 \frac{1 - (1/2X_1)\sin 2X_1}{X_2^2 + \cos^2 X_1}.$$

Here R' and β are functions of X_1, and hence of R, with a characteristic resonance denominator. The shapes of these functions are reminiscent of optical dispersion and absorption curves, respectively. The maximum in β occurs when $X_0 \cong (n + \tfrac{1}{2})\pi$ (n integer), the value of β being $2/[(n + \tfrac{1}{2})\pi\zeta]$ and the width of the peak at half-maximum $(n+\tfrac{1}{2})\pi\zeta$. The minimum value of β is about $(n+\tfrac{1}{2})\pi\zeta/2$. Figure 1 shows both magnitudes plotted as a function of X_0 for a value of $\zeta = 0.03$.

V. Comparison with Experimental Results

Figure 2a shows a profile presentation of the calculations of the neutron total cross-sections on the basis of the potential (3.1) with a depth $V_0 = 42$ MeV and a radius $R = 1.45 \times 10^{-13} A^{\frac{1}{3}}$ cm. The constant ζ is assumed to be 0.03 which corresponds to an absorption coefficient of $\kappa = 4.2 \times 10^{11}$ cm^{-1} in nuclear matter for neutrons of zero energy in free space. This means that the intensity of a beam of slow neutrons is reduced in nuclear matter to $1/e$ at a distance of $\kappa^{-1} = 2.4 \times 10^{-12}$ cm. The cross-sections are plotted as a function of the energy in units of $x^2 = (R/\lambda^2)$ and of the atomic weight. The letters denoting the maxima indicate the character of the resonance causing the maximum. Figure 3 contains a profile presentation of the observed cross-sections plotted against the same coordinates.

The experimental curves in Fig. 3 are averages over resonances. For higher A and small level distance this average was done by the measuring apparatus itself, for lower A the averaging was done in the drawing. The theoretical and the experimental curves do not extend to zero energy. They are broken off at an energy of about 50 keV. As is well known, the curves should go approximately as $\epsilon^{-\frac{1}{2}}$ at very low energy. The experimental curves are compiled from measurements by many workers.[19-25]

The comparison of these two figures shows that the theory can account for a number of striking features of the experimental results. In particular, the theory reproduces the drop of the cross-sections at low energies in the regions $A \sim 40$ and $100 < A < 140$.

It also reproduces the large cross-sections at low energy in the

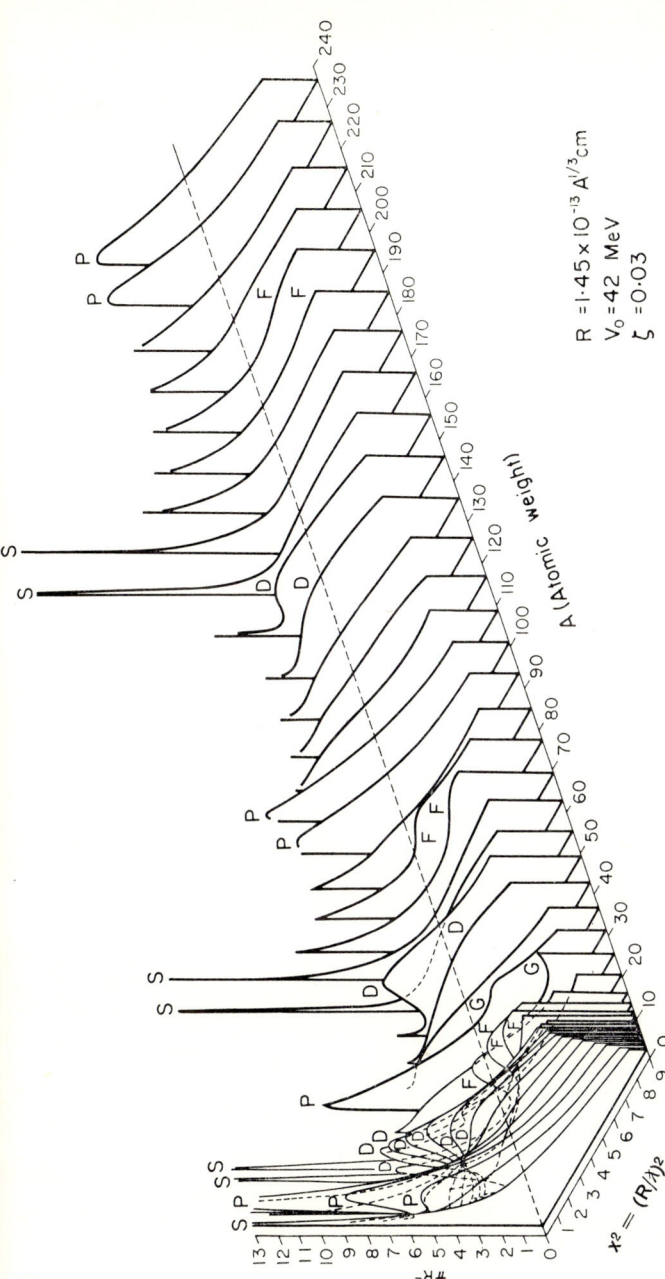

FIG. 2a. Calculated neutron total cross-sections as a function of energy and mass number, for a well depth $V_0 = 42$ MeV, radius $R = 1.45 \times 10^{-13} A^{1/3}$ cm, $\zeta = 0.03$. The energy ϵ is expressed in terms of
$$x^2 = [A^{2/3} \cdot A/10(A+1)]\epsilon,$$
where ϵ is in MeV.

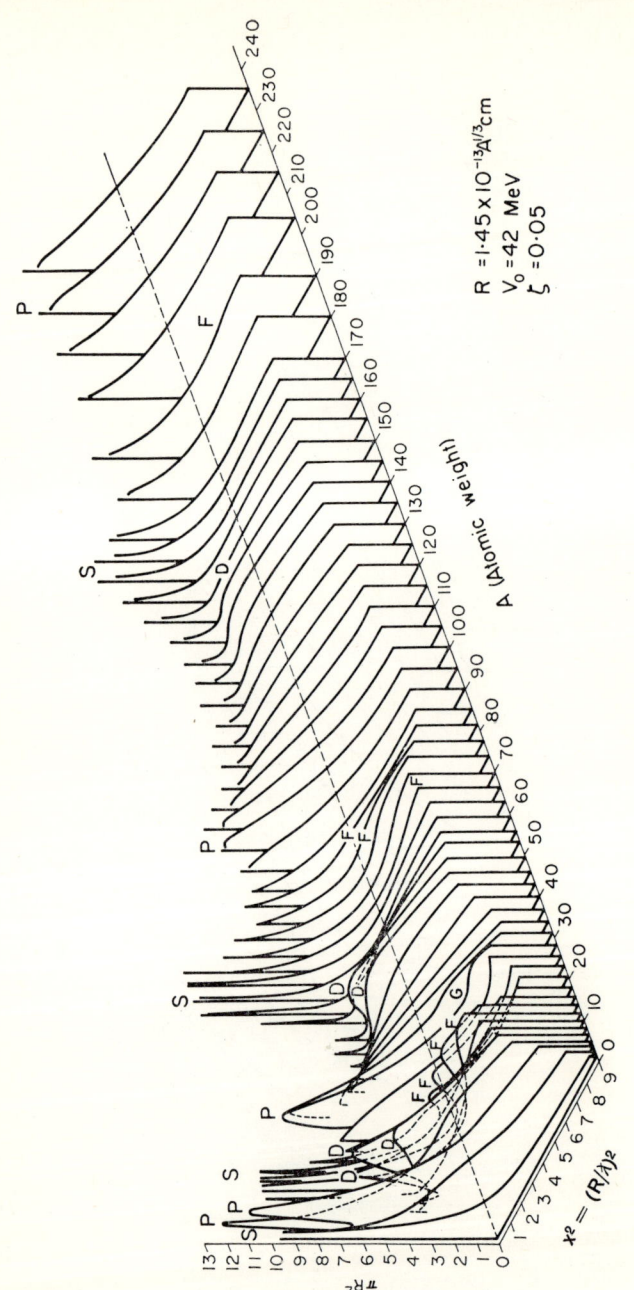

Fig. 2b. As Fig. 2a, but $\zeta = 0.05$.

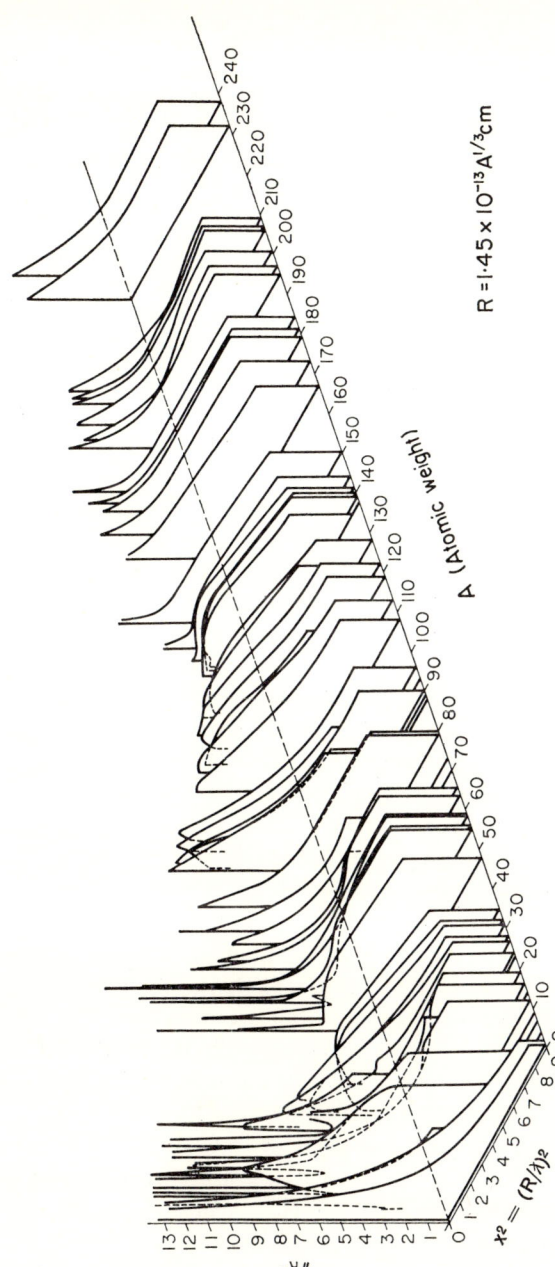

FIG. 3. Observed neutron total cross-sections as a function of energy and mass number. The energy is expressed in terms of x^2 as in Figs. 2a and 2b.

regions $A \sim 60$, $A \sim 90$, and $A \sim 150$. The large values at $A \sim 90$ are ascribed to a P resonance; whereas the other two regions are supposed to contain S resonances. P resonances are expected to fall off towards low energies; whereas the S resonances merge directly with the $(1/v)$ rise. The observed energy dependence indicates the P-resonance behavior in the region $A \sim 90$ and shows typical S-resonance behavior at $A \sim 60$ and 150. There is an indication of P-resonance at low energies for $A \sim 30$ as the theory predicts.

The theory also reproduces the type of maxima (D maxima) which are found for energies corresponding to $x^2 \sim 3$ in the regions $A \sim 40$, and $A \sim 140$. It seems that the predicted F-wave maximum near $A \sim 200$ is also observed. It is remarkable that one finds reasonably good agreement in the shape of the curves even at very low atomic numbers: $A < 20$.[26]

We are using here a different depth of the potential than in the calculations published previously by the same authors.[27] The previous calculations were based upon a well depth of only $V_0 = 19$ MeV. The change to $V_0 = 42$ MeV was suggested by Adair[28] and improves the agreement considerably. At the time of the first calculation only measurements for $A > 60$ were used. The similarity between the theoretical results for $V_0 = 42$ MeV and $V_0 = 19$ MeV for $A > 60$ can be explained as follows:

S-wave maxima at low energy occur if $RK_0 = r_0 A^{\frac{1}{3}} K_0 \cong (n + \tfrac{1}{2})\pi$ and P maxima if $r_0 A^{\frac{1}{3}} K_0 \cong n\pi$, where $K_0 = (2mV_0/\hbar^2)^{\frac{1}{2}}$. For $V_0 = 19$ MeV and $r_0 = 1.45 \times 10^{-13}$ cm, one gets therefore S maxima at $A \sim 38$ and 170, and P maxima at $A \sim 11$ and 90. For $V_0 = 42$ and the same r_0 one gets S maxima at $A \sim 11$, 55, and 150; P maxima at $A \sim 27$, 90, and 216. Hence, the P maximum near 90 and the S maximum near 160 are reproduced by both potential depths. The behavior of the curves in the neighborhood of these maxima also must be similar, in particular, the depression at low energy for values of A just below an S maximum. However, the experimental data for nuclei below $A = 60$ definitely indicate another S maximum near 55 and a strong low-energy depression for $A \sim 40$ as predicted by $V_0 = 42$. These features are not reproduced by the theoretical curves for $V_0 = 19$. We therefore believe that $V_0 = 42$ MeV yields a better model. It should be noted that the agreement is not very sensitive to a change of potential V_0 with a corresponding change of r_0 such that $V_0^{\frac{1}{2}} r_0$ stays constant.

The shapes of the total cross-section curves are quite sensitive to the value of the absorption constant ζ. An increase of ζ flattens the maxima and minima. Strong fluctuations in the calculated curves occur only at lower energies for values of $x < 1.5$. This is below 2 MeV at $A \sim 60$ and below 0.6 MeV at $A \sim 200$. At higher energies the contributions of the numerous angular momenta prevent the appearance of any pronounced maxima or minima. Therefore the determination of ζ by fitting the calculated curves to the experimental ones only gives the value of ζ for relatively low energies. We cannot exclude a change of ζ at energies of, say, more than 1 MeV or fluctuations in ζ from one value of A to another although below 1 MeV it seems that ζ cannot vary much as a function of A. In the low-energy region the determination is quite accurate. A change of ζ to 0.05 or to 0.02 would give rise to a worse agreement with experiments. Figure 2(b) shows the total cross-sections for $\zeta = 0.05$, and it is obvious that the maxima and minima are not as pronounced as in the experimental data.[29]

We now turn to the calculations of the angular distribution of the elastic scattering. Figure 4 shows the experimental results at 1 MeV as measured by Walt and Barschall.[30] The most characteristic features are the flat distributions around $A \sim 60$, a very strong forward peaking and a rise at backward angles at $A \sim 140$, and the appearance of a second maximum at 90° around $A \sim 180$. The calculation of the angular dependence (Fig. 5) is not unambiguous since the amount of compound elastic scattering is difficult to determine. Furthermore, the angular dependence of the compound elastic scattering depends upon the spin of the target nucleus. We therefore have shown in Fig. 5 the calculated angular distribution of the shape elastic scattering only. In Fig. 6 the compound elastic scattering is added in full which would correspond to the case in which the compound state decays exclusively via the entrance channel. The target spin was assumed to be zero. The actual

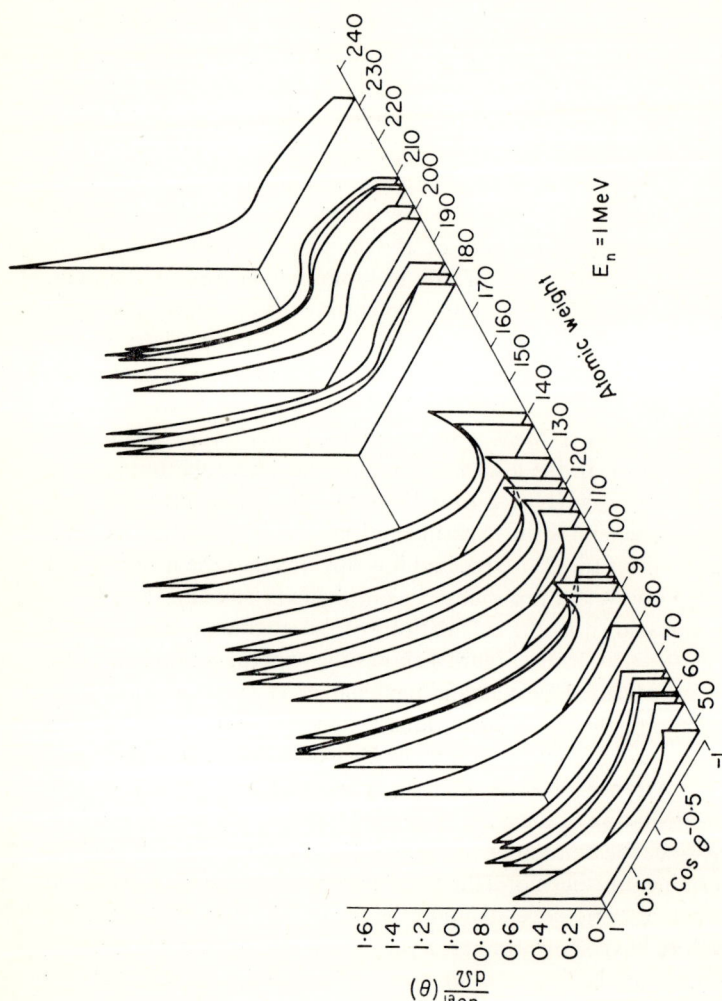

Fig. 4. Observed angular distribution (in barns/sterad) of the elastic 1-MeV neutron scattering as a function of $\cos\theta$ and the mass number A as measured by Walt and Barschall.

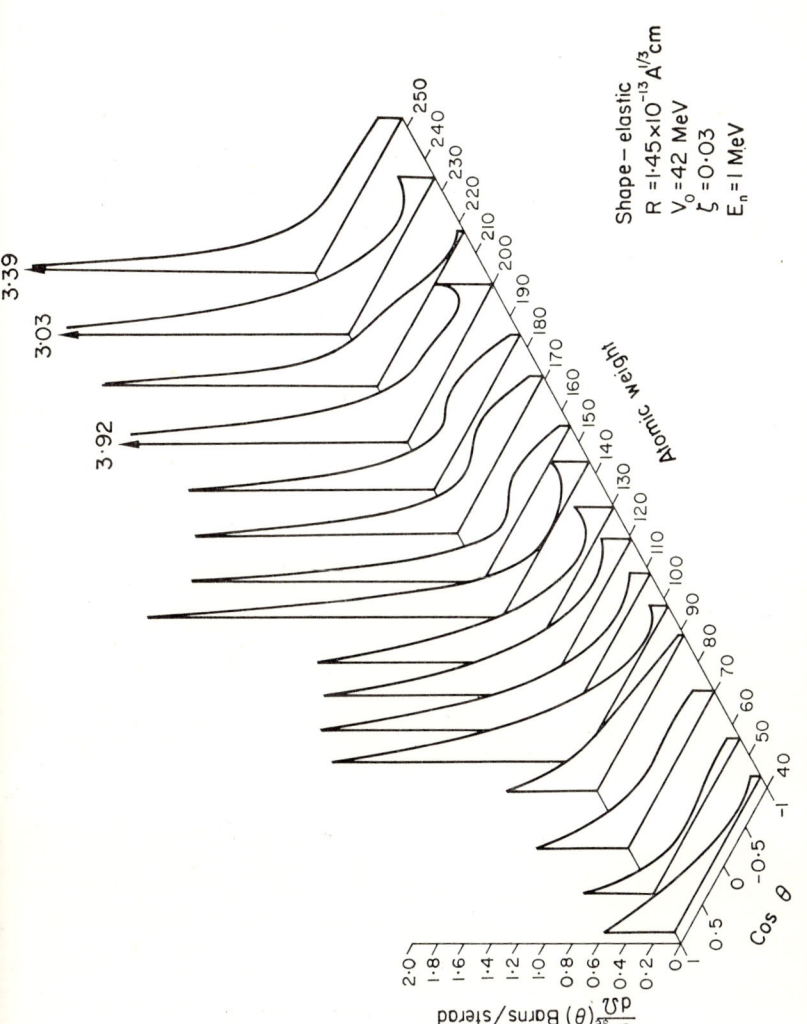

FIG. 5. Calculated angular distribution of the elastic neutron scattering (shape elastic only) as a function of $\cos\theta$ and the mass number A for a well $V_0 = 42$ MeV, $R = 1.45 A^{\frac{1}{3}} \times 10^{-13}$ cm, and $\zeta = 0.03$.

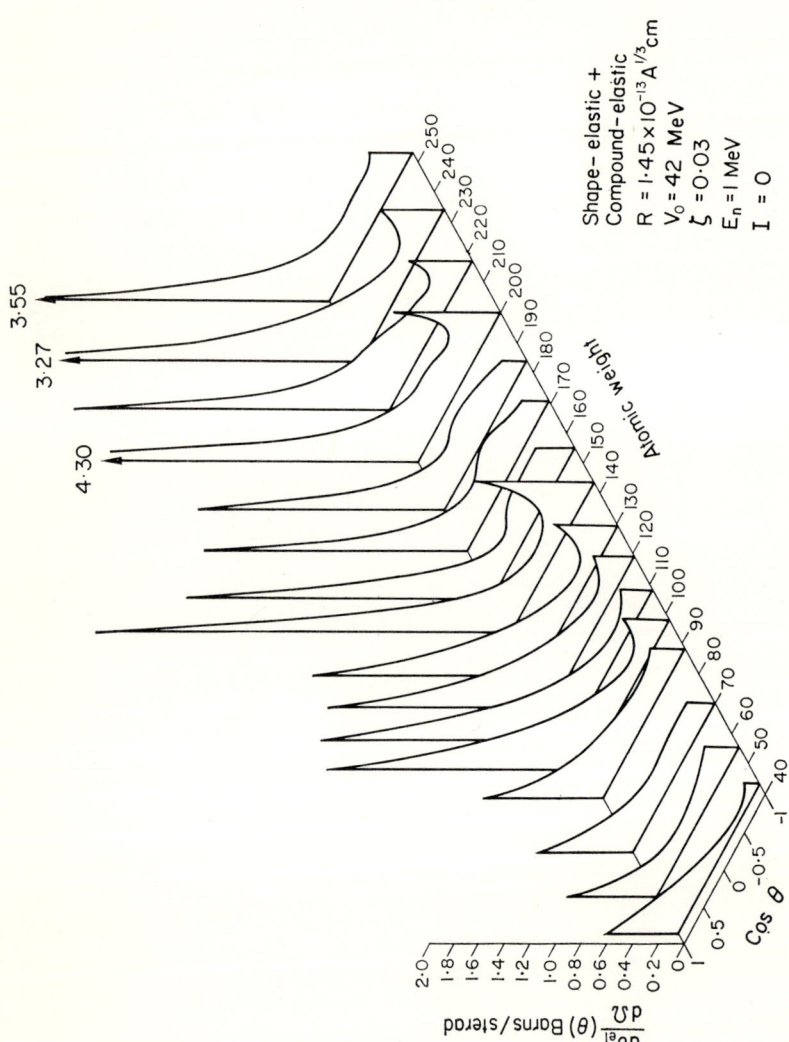

FIG. 6. Calculated angular distribution of the elastic neutron scattering (shape elastic plus maximum compound elastic) as a function of $\cos\theta$ and A for a well $V_0 = 42$ MeV, $R = 1.45 A^{\frac{1}{3}} \times 10^{-13}$ cm, and $\zeta = 0.03$.

MODEL FOR NUCLEAR REACTIONS WITH NUETRONS

$d\sigma_{el}/d\Omega$ must lie somewhere between Fig. 5 and Fig. 6. For nuclei with strong inelastic scattering, Fig. 5 should be the better approximation.

It is seen from Figs. 5 and 6 that some of the main features are again reproduced by the theory. The flatness of the distribution around $A \sim 60$ comes from the fact that the P contribution is very weak in this region and, at small angles, of opposite phase to the

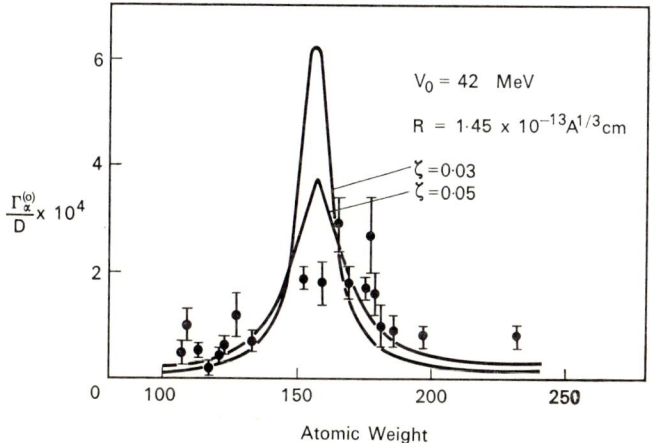

FIG. 7. Ratio $\Gamma_\alpha^{(0)}/D$ of neutron width to level distance for low energies as a function of A. Here $\Gamma_\alpha^{(0)} = \Gamma_\alpha(\epsilon^{(0)}/\epsilon)^{\frac{1}{2}}$ is the "reduced" width, and $\epsilon^{(0)}$ is taken to be 1 eV. The curves represent the calculated values for $\zeta = 0.03$ and 0.05. The points represent the observed values and the limits of error.

S scattering. This occurs always at values KR somewhat below a P resonance. The second maximum at 90° at high mass numbers is not too well reproduced. The theory shows it only between $A = 150$ and $A = 200$. The angular dependence above $A = 200$ does not seem to agree too well with the experiment. The angular distributions are not very sensitive to the choice of constants. The results with $V_0 = 19$ MeV are not very different from the ones shown here.

The agreement with experiments is also less satisfactory for the cross-section σ_c for the formation of the compound nucleus. It is difficult to measure σ_c directly since it includes the compound elastic scattering besides the reaction cross-section, and the former cannot easily be separated from the shape elastic scattering. At very low energies, however, the formation of the compound nucleus can be measured by studying the individual resonances (see section III). The relevant magnitude is the ratio Γ_α/D of the neutron width to the level distance, averaged over a number of neighboring resonances. The theoretical values Γ_α/D expected on the basis of $V_0 = 42$ MeV are shown in Fig. 7 together with a compilation[31] of the measurements[32-41] of Γ_α/D. Only recently has it been possible to measure the neutron widths of several resonances in one isotope, so that the average Γ_α and the level distance can be determined to some degree of reliability. It is seen that the expected maximum of Γ_α/D at $A \sim 155$ is noticeable, but it is not as strong as the theory predicts for the same value of ζ which gives the best fit for the total cross-sections ($\zeta = 0.03$). Also the values off peak are somewhat larger than predicted.

The fact that the resonance at $A \sim 155$ is not as strong as expected might be connected with the large deviations from sphericity which are ascribed to the nuclei in this region.[42] If the shape of the potential well is ellipsoidal, one would expect results which roughly represent averages over the spherical results taken over radii which lie between the smallest and the largest axis. This would give rise to a flattening of the maxima and a rise of the wings in the theoretical curves of Fig. 7.[43]

Although no direct measurement of the formation of the compound nucleus is possible, the measurements of inelastic cross-sections σ_{in} or reaction cross-sections σ_r can be used to compare with the theoretical predictions of σ_c. Evidently σ_{in} and σ_r must be smaller than σ_c. The difference $\sigma_c - \bar{\sigma}_r$ is the compound elastic cross-section which is expected to be rather small if inelastic scattering or other reactions are strong enough to compete for the decay of the compound state.

Walt and Barschall have determined inelastic scattering cross-sections σ_{in} at 1 MeV by subtracting the elastic scattering from the total scattering. The values of σ_{in} should be less than or equal to the theoretical values of σ_c.

Figure 8 shows a comparison between the observed inelastic cross-sections and the calculated σ_c at 1 MeV as functions of A. The observed values are of the expected order of magnitude, but they do not agree with the theoretical curve. The absence of the maximum at $A = 50$ might be explained by the fact that the compound elastic scattering is relatively high for these nuclei. The

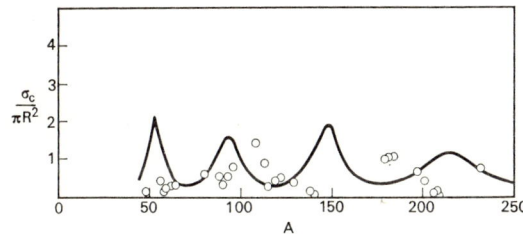

FIG. 8. The calculated cross-section σ_c for compound nucleus formation at $E_n = 1$ MeV and the observed reaction cross-section at 1 MeV as determined by Walt and Barschall. R is taken to be $1.45 \times 10^{-13} A^{\frac{1}{3}}$ cm. In the calculations the parameters V_0 and ζ were taken to be 42 MeV and 0.03, respectively.

same fact explains the low value of the inelastic cross-section in lead and bismuth. However, the expected maxima at $A = 90$ and 150 seem to occur at higher values of A. We have no explanation for these discrepancies.

There are many measurements of inelastic cross-sections at somewhat higher energies. They all indicate that the values are not too far from $\pi(R + \lambda)^2$, which is the value one would expect if the neutron wave were totally absorbed in contrast to our findings of $\zeta \sim 0\cdot03$. Especially the measurements at 14 MeV[3] indicate this fact. On the basis of this evidence one would conclude that the value of ζ is strongly energy dependent and reaches a value ($\zeta \gtrsim 0.12$) corresponding to almost total absorption in a medium-sized nucleus certainly at 14 MeV but most likely already at energies as low as 4.5 MeV. The latter conclusion is based upon measurements of inelastic cross-sections by Lonsjo, Taylor, and Bonner.[44] In this connection it is interesting that the calculations of Morrison, Muirhead, and Rosser[45] also give a very strong increase with energy of the absorption of nucleons in nuclear matter just in the region which corresponds to incident neutrons of 1 MeV. These calculations are based on the Goldberger method[46] of the scattering of free particles with the application of the Pauli principle. The effect of the exclusion principle alone causes a sharp drop of the energy exchange with decreasing energy.[47]

It is apparent that our model is much less successful in reproducing the strength of compound nucleus formation than in reproducing the total and elastic scattering. It gives too much variation with A of Γ_α/D at low energies and probably too little compound nucleus production at 1 MeV and higher, although it is possible to explain the discrepancies at higher energy by assuming that ζ increases with energy above 1 MeV.

The discrepancies may come from two possible sources: (A) The potential $V(r)$ as given by (3.1) may not be the shape best fitted for the model. (B) The attempt of this paper, the description of the gross behavior of a nucleus by a complex one-particle potential, may be unsuccessful. In connection with (A) it must be noted that the potential (3.1) necessarily is an oversimplified version, since it is physically impossible that the potential well actually has a discontinuity in the form of a sudden jump at $r = R$. It might be that a rounding-off of the corners of the potential well will improve the agreement with experiments.

The smoothing of the edges of the square-well potential was of significance for the interpretation of the elastic proton scattering with heavy nuclei. This scattering has been measured with protons of an energy of about 18 MeV by Gugelot,[48] Burkig and Wright,[49] and by Cohen and Neidigh.[50] The results cannot be interpreted on the basis of a potential (3.1) with sharp edge, as shown by Chase and Rohrlich.[51] However, Woods and Saxon[52] have shown recently that a rounding-off even within the small interval of 0.5×10^{-13}

cm changes the results considerably and brings them into much better agreement with the experiments.

It is possible, therefore, that the smoothing of the discontinuity of V at $r = R$ would also improve the agreement of theory and experiment in respect to compound nucleus formation. It would decrease the reflection of the neutron wave at the nuclear surface and hence increase the cross-section σ_c when all other constants (V_0, R, ζ) are unchanged. It remains to be seen whether the rounding-off of the potential well improves the agreement with respect to σ_c and Γ_α/D and with respect to the angular distribution of the elastic cross-section, without destroying the agreement of the total and elastic cross-sections.

Calculations are under way to investigate these possibilities.

It must be pointed out that one should never expect any exact agreement between the predictions based upon a model of this type and the observed cross-sections. The very nature of this attempt to describe a complicated many-body problem by a simple one-body potential implies that the model can only contain the main features of the situation. Apart from this general limitation it should be kept in mind that we have used here a potential which has a particularly simple dependence on the radius and on the mass number. We have assumed the same radial dependence for the real and imaginary part which is very probably too strict an assumption. We have neglected spin-dependent forces as observed by Adair and co-workers,[53] and we have excluded any special features connected with the shell structure.

The purpose of the proposed approach is to connect some characteristic salient features of the nuclear cross-sections with simple nuclear properties rather than to construct a theory which will produce the exact quantitative details of the observations.

Acknowledgments

We are greatly indebted to many experimental physicists who in the course of the last few years have discussed with us their results before publication and have helped us adjust our models to the latest values. In particular, we owe special gratitude to H. H. Barschall, M. Walt, and collaborators at Wisconsin; R. K. Adair, J. A. Harvey, D. J. Hughes at Brookhaven; N. Nereson of Los Alamos; and T. W. Bonner of the Rice Institute for their invaluable help and advice and readiness to furnish their newest results.

We also wish to express our gratitude to those who have participated in the important, but tedious, work of computing. Our thanks are offered to Fern Abrams, Harvey Amster, Betty Campbell, Elgie Ginsburgh, Dr. A. Glassgold, Mida Karakashian, Barbara Levine, Edith Moss, Hannah Paul, Evelyn Walker, and Hannah Wasserman at the Massachusetts Institute of Technology; Elaine Scheer at Nuclear Development Associates, Inc.; and Jane Levin and Phyllis Levy at Brookhaven National Laboratory.

One of us (C.E.P.) wishes to acknowledge the helpful cooperation he received at Brookhaven National Laboratory during the final stage of this work prior to publication.

Notes

[1] H. Bethe, *Phys. Rev.* **57**, 1125 (1940); Feshbach, Peaslee, and Weisskopf, *Phys. Rev.* **71**, 145 (1947); H. Feshbach and V. F. Weisskopf, *Phys. Rev.* **76**, 1550 (1949); E. P. Wigner, *Phys. Rev.* **73**, 1002 (1948); E. P. Wigner, *Am. J. Phys.* **17**, 99 (1949).

[2] H. Feshbach and V. F. Weisskopf, *Phys. Rev.* **76**, 1550 (1949).

[3] Phillips, Davies, and Graves, *Phys. Rev.* **88**, 600 (1952).

[4] This terminology will become obvious later on. B. T. Feld [*Experimental Nuclear Physics*, edited by E. Segrè (John Wiley and Sons, Inc., New York, 1953), Vol. 2] calls σ_{ce} the "capture elastic" cross-section.

[5] The appearance of the length R' is a consequence of our general treatment of the nuclear resonance in Appendix 1. In the special derivation of the Breit–Wigner formula, as given in Feshbach, Peaslee, and Weisskopf (footnote 1) or J. M. Blatt and V. F. Weisskopf, *Theoretical Nuclear Physics* (John Wiley and Sons, Inc., New York, 1952), assumptions are made which make R' constant and equal to the nuclear radius R. It is shown in this paper that these assumptions probably are valid only in special cases as in the case of strong coupling, for example.

[6] H. Bethe, *Phys. Rev.* **57**, 1125 (1940).

[7] Fernbach, Serber, and Taylor, *Phys. Rev.* **75**, 1352 (1949).

[8] K. W. Ford and D. Bohm, *Phys. Rev.* **79**, 745 (1950).

[9] This follows the notation of Morse, Lowan, Feshbach, and Lax, U.S. Navy Department of Research and Inventions Report No. 61.1R, 1945 (unpublished).

[10] W. Hauser and H. Feshbach, *Phys. Rev.* **87**, 366 (1952).

[11] L. Wolfenstein, *Phys. Rev.* **82**, 690 (1951).
[12] G. Racah, *Phys. Rev.* **61**, 186 (1942): **62**, 438 (1942).
[13] J. M. Blatt and L. C. Biedenharn, *Revs. Modern Phys.* **24**, 258 (1952).
[14] Biedenharn, Blatt, and Rose, *Revs. Modern Phys.* **24**, 249 (1952).
[15] L. C. Biedenharn, Oak Ridge National Laboratory Report ORNL-1501, May 28, 1953 (unpublished).
[16] It is worth while noting that eq. (3.15) may be derived from the general analysis of Blatt and Biedenharn by combining the definitions of average cross-sections as given in section II and the statistical assumptions. The chief elements of the latter are (1) nonoverlap of resonances and (2) random phases for the scattering matrix so that, upon averaging over possible ways of forming the compound nucleus, interference terms average to zero.
[17] From here on we use the symbol Γ_α for "neutron width" since the entrance channel α is a *neutron* channel in all cases which we treat in this paper.
[18] It is plausible to assume that, in the case of strong coupling, the imaginary part of the potential is of the same order as the real part. An imaginary part that is much larger than the real one would imply that the absorption takes place over distances small compared to the wavelength in the interior. Hence X_1' is about twice as large as X_0 which leads to a Γ_α/D half as large as in Feshbach, Peaslee, and Weisskopf. This strong coupling result is somewhat more consistent than the result in Feshbach, Peaslee, and Weisskopf. In the latter paper the boundary condition was chosen such that the wave inside the nucleus is a sine wave $\sin(Kr - \delta)$, an assumption that is contrary to the idea of strong compound nucleus formation. In fact, a wave $\exp(+\Omega r)\sin(Kr-\delta)$ with $\Omega \sim K$ would be more consistent and does lead to the same result as the one above.
[19] H. H. Barschall, *Phys. Rev.* **86**, 431 (1952).
[20] Miller, Adair, Bockelman, and Darden, *Phys. Rev.* **88**, 83 (1952).
[21] Walt, Becker, Okazaki, and Fields, *Phys. Rev.* **89**, 1271 (1953).
[22] Okazaki, Darden, and Walton, *Phys. Rev.* **93**, 461 (1954).
[23] N. Nereson and S. Darden, *Phys. Rev.* **89**, 775 (1953); *Phys. Rev.* **94**, 1678 (1954); and unpublished data on Li and B (private communication).
[24] C. F. Cook and T. W. Bonner, *Phys. Rev.* **94**, 651 (1954); McCrary, Taylor, and Bonner, unpublished data on Li (private communication).
[25] *Neutron Cross Sections*, U.S. Atomic Energy Commission Report AECU-2040 (Technical Information Division, Department of Commerce, Washington, D.C., 1952), and three supplements (unpublished).
[26] See also C. E. Porter, *Bull. Am. Phys. Soc.* **29**, No. 5, 25 (1954).
[27] Feshbach, Porter and Weisskopf, *Phys. Rev.* **90**, 166 (1953).
[28] R. K. Adair, *Phys. Rev.* **94**, 737 (1954).
[29] The disagreement is worse for high values of A. This might be an indication of a slight decrease of ζ with the mass number. If the absorption were concentrated in a surface layer of given thickness, one would expect a similar effect [see M. H. Johnson and E. Teller, *Phys. Rev.* **93**, 357 (1954)].
[30] M. Walt and H. H. Barschall, *Phys. Rev.* **93**, 1062 (1954).
[31] R. S. Carter *et al.*, *Phys. Rev.* (to be published).
[32] F. G. P. Seidl, Hughes, Palevsky, Levin, Kato, and Sjöstrand, *Phys. Rev.* **95**, 476 (1954); and private communication.

[33] R. S. Carter and J. A. Harvey, *Phys. Rev.* **95**, 645(A) (1954).

[34] Foote, Landon, and Sailor, *Phys. Rev.* **92**, 656 (1953).

[35] Sailor, Landon, and Foote, *Phys. Rev.* **93**, 1292 (1954).

[36] Pilcher, Carter, and Stolovy, *Phys. Rev.* **95**, 645(A) (1954); and private communication.

[37] Hughes, Kato, and Levin, *Phys. Res.* **92**, 1094 (1953); and private communication.

[38] R. L. Christensen, *Phys. Rev.* **92**, 1509 (1953).

[39] Melkonian, Havens, and Rainwater, *Phys. Rev.* **92**, 702 (1953).

[40] L. Bollinger, unpublished data on Sb (private communication). We wish to thank Dr. Bollinger for making his results available in advance of publication.

[41] V. E. Pilcher and R. S. Carter (private communication).

[42] A. Bohr and B. R. Mottelson, *Kgl. Danske Videnskab. Selenskab Mat. fys. Medd.* **27**, 16 (1953).

[43] This thought was suggested to us by A. Bohr and B. R. Mottelson.

[44] Lonsjo, Taylor, and Bonner (private communication). We are grateful to the authors for showing us their results before publication.

[45] Morrison, Muirhead, and Rosser, *Phil. Mag.* **44**, 1326 (1953).

[46] M. Goldberger, *Phys. Res.* **74**, 1269 (1948).

[47] V. F. Weisskopf, *Science* **113**, 101 (1951).

[48] P. C. Gugelot, *Phys. Rev.* **87**, 525 (1952).

[49] J. W. Burkig and B. T. Wright, *Phys. Rev.* **82**, 451 (1951).

[50] B. L. Cohen and R. V. Neidigh, *Phys. Rev.* **93**, 282 (1954).

[51] D. M. Chase and F. Rohrlich, *Phys. Rev.* **94**, 81 (1954).

[52] R. D. Woods and D. S. Saxon (private communication). We are grateful to the authors for showing us their results before publication.

[53] Darden, Field, and Adair, *Phys. Rev.* **93**, 931 (1953).

[54] At first sight this seems puzzling since $\sigma_r = \Sigma_\beta \sigma_{\alpha\beta}$. It must be remembered that eq. (2.3a) uses the diagonal element of S; whereas (A.18) uses off-diagonal elements. The connection between these elements is established by the unitary nature of S: $1 - |S_{\alpha\alpha}|^2 = \Sigma_{\beta \neq \alpha}(S_{\alpha\beta})^2$. In order to insure the validity of this equation up to the order g^2, one must include the last two terms of (A.14) in $S_{\alpha\alpha}$, but it is not necessary to include them in $S_{\alpha\beta}'$.

[55] E. P. Wigner and L. Eisenbud, *Phys. Rev.* **72**, 29 (1947).

Appendix 1

The scattering amplitude at low energies

The scattering amplitude η is the diagonal element $S_{\alpha\alpha}$, of the scattering matrix, where the index α refers to the entrance channel. The matrix S is given by the following expression [see Blatt and Weisskopf, formula (X, 4.11)]:

$$S_{\alpha\beta} = \exp[-i(k_\alpha + k_\beta)R]S_{\alpha\beta}'. \tag{A.1}$$

Here R is the nuclear radius, and

$$S' = (1 + i\mathscr{R}')/(1 - i\mathscr{R}'), \tag{A.1a}$$

MODEL FOR NUCLEAR REACTIONS WITH NEUTRONS 267

where $\mathscr{R}'$ is connected with the derivative matrix $\mathscr{R}$ by

$$\mathscr{R}_{\alpha\beta}' = (k_\alpha k_\beta)^{\frac{1}{2}} \mathscr{R}_{\alpha\beta},$$

and k_α, k_β are the channel wave numbers at the energy E. The matrix $\mathscr{R}$ is defined on page 545 of Blatt and Weisskopf. It can be expressed in the following form [see sec. X (4.22)]:

$$\mathscr{R}_{\alpha\beta} = \sum_s \frac{y_{s\alpha} y_{s\beta}}{E_s - E}, \tag{A.2}$$

where E_s are the resonance energies and the $y_{s\alpha}$ are magnitudes which are connected with the channel widths Γ_α^s (partial widths for the decay via the channel α):

$$\Gamma_\alpha^s = 2k_\alpha y_{s\alpha}^2. \tag{A.3}$$

Each $y_{s\alpha}$ is real but its sign might be positive or negative. We make the reasonable assumption that the signs are distributed at random.

In what follows we will assume that we work in an energy region for which, first, $k_\alpha R \ll 1$, and, second, $\Gamma^s \ll D_s$, where D_s is the interval

$$D_s = \tfrac{1}{2}(E_{s+1} - E_{s-1}),$$

which includes the resonance E_s from mid-point to mid-point. We also assume that the values Γ_α^s and D^s have the same order of magnitude for all resonances in an energy interval I which includes many resonances but which is small compared to energy intervals occurring in single-particle problems (say $I \sim 10$ keV for heavy nuclei).

Let us surround each resonance E_s by an energy interval D_s from $\tfrac{1}{2}(E_{s-1} + E_s)$ to $\tfrac{1}{2}(E_s + E_{s+1})$. In this interval we can write the matrix $R_{\alpha\beta}'$ in the form

$$\mathscr{R}_{\alpha\beta}' = \frac{y_{s\alpha}' y_{s\beta}'}{E_s - E} + g_{\alpha\beta}, \tag{A.4}$$

where $y_{s\alpha}' = (k_\alpha)^{\frac{1}{2}} y_{s\alpha}$. The matrix $g_{\alpha\beta}$ has no singularities in D_s and is given by

$$g_{\alpha\beta} = \sum_{t \neq s} \frac{y_{t\alpha}' y_{t\beta}'}{E_t - E}. \tag{A.4a}$$

We now estimate the order of magnitude of $g_{\alpha\beta}$. Here it is important to take into account that the signs of the $y_{t\alpha}'$ are distributed at random over the different resonances t and the different channels α.

First we split $g_{\alpha\beta}$ into the contributions of neighboring and far-off levels:

$$\begin{aligned} g_{\alpha\beta} &= g_{\alpha\beta}' + r_{\alpha\beta}, \\ g_{\alpha\beta}' &= \sum_{t \neq s}{}' \frac{y_{t\alpha}' y_{t\beta}'}{E_t - E}, \\ r_{\alpha\beta} &= \sum_{t \neq s}{}'' \frac{y_{t\alpha}' y_{t\beta}'}{E_t - E}, \end{aligned} \tag{A.4b}$$

where the prime on the summation sign means that the sum should be extended only over resonances within the interval I, and the double prime means extension over the resonances outside I. Because of the random signs of $y_{t\alpha}'$, the terms in the sums (A.4b) have random signs for $\alpha \neq \beta$ and only the immediate-neighbor resonances contribute appreciably to $g_{\alpha\beta}$; on the same grounds $r_{\alpha\beta}$ can be neglected. We, then, obtain the estimate:

$$|g_{\alpha\beta}'| \cong |g_{\alpha\beta}| \sim (\Gamma_\alpha \Gamma_\beta)^{\frac{1}{2}}/D, \quad r_{\alpha\beta} \approx 0, \quad \alpha \neq \beta. \tag{A.5}$$

We understand by Γ_α (without superscript) the average value of Γ_α^s in the interval I. For $\alpha = \beta$, all terms have the same sign for $E_t > E$ or $E_t < E$. Since there are roughly an equal number of levels above and below E_s in the interval I, we get the following estimate:

$$g_{\alpha\alpha}' \sim \Gamma_\alpha/D. \tag{A.5a}$$

The order of magnitude of the contribution $r_{\alpha\alpha}$ from the faraway levels is quite undetermined. However, the scattering cross-section between resonances turns out to be $4\pi R^2(1 + r_{\alpha\alpha}/x)^2$. Experimentally, we know that this cross-section is of the order of $4\pi R^2$, and hence we conclude $r_{\alpha\alpha} \sim k_\alpha R \equiv x$.

It follows from (A.5) and (A.5a) that the $g_{\alpha\beta}$ are all small compared to unity in the energy region considered, and we proceed to expand (A.1) in powers of g. For this purpose we introduce the *factorable* matrix

$$T_{\alpha\beta} = -y_{s\alpha}' y_{s\beta}'/\Delta, \quad \Delta = E - E_s. \tag{A.6}$$

The following relation holds:

$$T^n = B^{n-1}T, \quad n \geq 1,$$
$$B = \Sigma_\beta T_{\beta\beta}. \tag{A.7}$$

The quantity B is a number and it is connected with the total width. $\Gamma^s = \Sigma_\beta \Gamma_\beta^s$ of the resonance s:

$$B = -\tfrac{1}{2}\Gamma^s/\Delta. \tag{A.8}$$

Hence, if a matrix is a function of T which can be expressed as a power series with *powers larger than zero*

$$A(T) = \sum_{n=1}^{\infty} a_n T^n,$$

we get the relation

$$A_{\alpha\beta} = B^{-1} A(B) T_{\alpha\beta}. \tag{A.9}$$

We may now expand (A.1), noting that $\mathscr{R}' = T + g$:

$$S' = -1 + \frac{2}{1 - i(T + g)}$$
$$= -1 + \frac{2}{1 - iT} + \frac{1}{1 - iT}(2ig)\frac{1}{1 - iT}$$
$$\qquad - \frac{2}{1 - iT} g \frac{1}{1 - iT} g \frac{1}{1 - iT} + \cdots.$$

From (A.9) we have $1/(1 - iT) = 1 + T/(1 - iB)$, so that

$$S' = 1 + \frac{2iT}{1 - iB} + 2ig - 2g^2 - 2\frac{(gT + Tg)}{1 - iB} + S^*,$$

$$S^* = -2i\frac{TgT}{(1 - iB)^2} - 2i\frac{Tg^2 + gTg + g^2T}{(1 - iB)}$$
$$- 2\frac{TgTg + Tg^2T + gTgT}{(1 - iB)^2} + 2i\frac{TgTgT}{(1 - iB)^3}. \tag{A.10}$$

The scattering amplitude η_0 is the diagonal element $S_{\alpha\alpha}$. We first note the cross-sections which would follow from (A.10) when we put $g = 0$: ($x = k_\alpha R$)

$$\eta_0 = e^{-2ix}\left(1 + \frac{2iT_{\alpha\alpha}}{1 - iB}\right) = e^{-2ix}\left(1 - \frac{i\Gamma_\alpha^s}{\Delta + i\Gamma^s/2}\right), \tag{A.11}$$

and hence we get the well-known expressions:

$$\sigma_{el}^{(0)} = \pi\lambda^2 \left|2x + \frac{\Gamma_\alpha^s}{\Delta + i\Gamma^s/2}\right|^2,$$
$$\sigma_r^{(0)} = \pi\lambda^2 \frac{\Gamma_\alpha^s(\Gamma^s - \Gamma_\alpha^s)}{\Delta^2 + (\Gamma^s/2)^2}. \tag{A.12}$$

We now proceed to neglect all terms in the expansion (A.10) which would give rise to terms in σ_{el} and σ_r of the order of $\delta\sigma_r$ and $\delta\sigma_{el}$, respectively, with

$$\delta\sigma_r = \sigma_c^{(0)}f, \quad \delta\sigma_{el} = \sigma_c^{(0)}f, \tag{A.13a}$$

where $\sigma_c^{(0)} = \pi\lambda^2\Gamma_\alpha\Gamma/[\Delta^2 + (\Gamma^2/4)]$ and f is a small number,

$$f \sim x \quad \text{or} \quad f \sim \Gamma/D. \tag{A.13b}$$

This means we will neglect cross-sections which are small by a factor Γ/D or x, compared to $\sigma_c^{(0)}$, which can be regarded as the one-level value of the cross-section for the formation of the compound nucleus.

It will be shown below that all terms of (A.10) contained in S^* give rise to corrections in the cross-sections of the order (A.13) or smaller. Hence, within the accuracy (A.13), we can write S' in the form:

$$S' = 1 + \frac{2iT}{1 - iB} + 2ig - 2g^2 - \frac{2(gT + Tg)}{1 - iB}. \tag{A.14}$$

To prove this point, we first examine the effect of a small addition $\Delta\eta'$ to $\eta' = S_{\alpha\alpha}'$ on the cross-sections: We get

$$\Delta\sigma_{el} = \pi\lambda^2 2\,\text{Re}[(e^{2ix} - \eta')\Delta\eta'^*] \lesssim 4\pi\lambda^2 |\Delta\eta'|,$$
$$\Delta\sigma_r = \pi\lambda^2 2\,\text{Re}[\eta'^*\Delta\eta'] \leq 2\pi\lambda^2 |\Delta\eta'|, \tag{A.15}$$

where $\text{Re}(a)$ is the real part of a. The former relation follows from the fact that $|e^{2ix} - \eta'| \leq 2$ and the latter relation from the fact that $|\eta'| \leq 1$. As an example, we discuss the omission of the first term in the expression S^* in (A.10). We find according to (A.5) and (A.5a)

$$|\Delta\eta'| \sim |(TgT)_{\alpha\alpha}/(1-iB)^2| \approx \frac{\Gamma_\alpha \Gamma}{\Delta^2 + (\Gamma^s/2)^2}[(\Gamma/D) + x],$$

and hence the contributions to the cross-section of this term are negligible according to (A.15) and (A.13). Similar considerations show that the other neglected terms in (A.10) contribute the same or less to the cross-sections.

We now single out the diagonal element of S' because of its significance for the scattering amplitude. We can write $S_{\alpha\alpha'}$ from (A.14) and (A.13b) in the following form:

$$S_{\alpha\alpha'} \cong \exp(2ir_{\alpha\alpha})\left[\left(1 + \frac{2iT_{\alpha\alpha}}{1-iB}\right) + 2ig_{\alpha\alpha'} - 2(g'^2)_{\alpha\alpha} - \frac{4(g'T)_{\alpha\alpha}}{1-iB}\right].$$

This expression differs from the diagonal element of (A.14) by terms which are of the order S^* and therefore negligible, as, for example, $r_{\alpha\alpha}(g'T)_{\alpha\alpha}/(1-iB)$. We can write in it the form

$$S_{\alpha\alpha'} = \exp(2ir_{\alpha\alpha}) \times \left[1 + \frac{2iT_{\alpha\alpha}}{1-iB} + \frac{T_{\alpha\alpha}}{1-iB}G_1 + iG_2 + G_3\right], \quad (A.14a)$$

or, according to (A.1)

$$S_{\alpha\alpha} = \eta_0 = \exp(-2ik_\alpha R') \times \left[1 + \frac{2iT_{\alpha\alpha}}{1-iB} + \frac{T_{\alpha\alpha}}{1-iB}G_1 + iG_2 + G_3\right], \quad (A.14b)$$

where

$$R' = R - r_{\alpha\alpha}/k_\alpha. \quad (A.16)$$

$$G_1 = 4 \sum_{t \neq s}{}' \sum_\beta \frac{y_{t\alpha'} y_{t\beta'}(y_{s\beta'}/y_{s\alpha'})}{E_t - E},$$

$$G_2 = 2g_{\alpha\alpha'}, \quad (A.17)$$

$$G_3 = -2(g'^2)_{\alpha\alpha}.$$

This is the form which is used in the text. The orders of magnitude of these real functions are given by (2.11).

The following simplification can be used if one calculates the scattering cross-section σ_{el} and the transfer-cross-sections $\sigma_{\alpha\beta}$ (cross-section of the reaction $\alpha \to \beta$):

$$\sigma_{\alpha\beta}{}^{(0)} = \pi\lambda^2 |S_{\alpha\beta'}|^2 \quad (A.18)$$

within the limits of accuracy (A.13). It turns out that the last two terms in (A.14) give rise to nonnegligible contributions only to σ_r when expression (2.3a)

MODEL FOR NUCLEAR REACTION WITH NEUTRONS   271

is used. In expression (2.3) for σ_{el} and in expression (A.18) for $\sigma_{\alpha\beta}$, the two last terms of (A.14) give rise to contributions which can be neglected according to (A.13).[54] Hence for the calculation of $\sigma_{\alpha\beta}^{(0)}$ and $\sigma_{el}^{(0)}$ we may use the shorter form

$$S' = 1 + [2iT/(1 - iB)] + 2ig,$$

or

$$S_{\alpha\beta}' \simeq \delta_{\alpha\beta} \exp(2ir_{\alpha\alpha}) + 2i \sum_r{}' \frac{y_{r\alpha}' y_{r\beta}'}{E - E_r + i\Gamma^r/2}, \qquad (A.19)$$

where the sum is extended over all resonances within the interval I. Actually the imaginary part $+i\Gamma^r/2$ in the denominator of (A.16) should be found only in the term $r = s$, but the addition in the other terms leads only to errors smaller than (A.13). We then get for the cross-section $\sigma_{\alpha\beta}$

$$\sigma_{\alpha\beta} = \pi\lambda^2 \, | \, S_{\alpha\beta}' \, |^2 = 4\pi\lambda^2 \left| \sum_r{}' \frac{y_{r\alpha}' y_{r\beta}'}{E - E_r + i\Gamma^r/2} \right|^2, \qquad (A.20)$$

and for the scattering cross-section

$$\sigma_{el} = \pi\lambda^2 \, | \, e^{2ix} - S_{\alpha\alpha}' \, |^2,$$
$$= \pi\lambda^2 \left| \exp(2ik_\alpha R') - 1 + \sum_r \frac{i\Gamma_\alpha^r}{E - E_r + i\Gamma^r/2} \right|^2. \qquad (A.21)$$

According to (A.19), the value of $\sigma_{\alpha\beta}$ goes to zero between two resonances E_s and E_{s+1} if the sign of $y_{s\alpha}' y_{s\beta}'$ is the same for both resonances. If the sign is opposite, no zero occurs. Note that this statement is good only to the accuracy (A.13). At the zero of (A.17) the actual cross-section might still be of the order (A.13).

We note in (A.21) that the potential scattering amplitude $\exp(2ikR') - 1$ corresponds to the scattering by an impenetrable sphere of radius R' as given by (A.16). The quantity R' itself is a function of the energy, which is slowly varying and changes only over intervals much larger than D.

The forms (A.15) and (A.16) correspond to the Breit–Wigner formulas used in the literature before the more exact investigations by Wigner and Eisenbud.[55] The amplitudes contain characteristic sums over the contributions of the different resonances with the imaginary contribution $i\Gamma^r/2$ in the denominator. It has been pointed out repeatedly that the forms (A.17) and (A.18) are not exactly correct. We have shown, however, that they are valid within the errors given by (A.13).

9

Nuclear Reactions at High Energies†

R. SERBER

THE general features of the high-energy nuclear reactions which have been observed at the Radiation Laboratory can be understood in terms of a picture which is in its main outlines quite simple, though quite different from the description appropriate at lower energies. In trying to understand what happens when a nucleus is bombarded by a high-energy neutron or proton, the first consideration that comes to mind is that the collision time between the incident particle and a particle in the nucleus is short compared to the time between collisions of the particles in the nucleus. This suggests that the first step in the process can be regarded in terms of collisions between the incident particle and the individual nuclear particles. We are thus led to ask the properties of the high-energy scattering between free nucleons. There are two salient points. First, the total cross-section for scattering of one nucleon by another is inversely proportional to the energy of the incident particle. The mean free path of a nucleon traversing nuclear matter increases with its energy; at sufficiently high energies the nucleus begins to be transparent to the bombarding particles. Secondly, the incident particle loses only a small fraction of its energy to the struck one. The momentum transfer, which is nearly perpendicular to the direction of the incident particle, is of the order $\hbar/a$, where a is the range of nuclear forces, and does not increase with increasing energy. In case of an exchange collision, we continue to call the high-energy emergent particle the incident one, even though it has changed its charge.

† *Phys. Rev.* **72**, 1114 (1947).

Since the momentum transfer, $\hbar/a$, is not large compared to the characteristic momentum, $\hbar/d$, of particles in the nucleus (with d the mean separation of nuclear particles), it is not true that the collisions made by the incident particle can be considered as collisions between free particles; interference between particles in the nucleus can be important. Such an interference effect can be expected because of the degeneracy of nuclear matter. If nuclear matter is represented as a degenerate Fermi gas, it is clear that collisions with small momentum transfers will be discouraged, since these tend to lead from an occupied state to another already occupied. An estimate of this effect indicates that as a result the mean free path of a high-energy particle ($\sim$ 100 MeV) traversing nuclear matter will be increased over what would be expected for collisions between free particles by a factor of about 5/3. The mean kinetic energy transfer to the struck particle per collision is increased in the same ratio.

We estimate that the mean free path for a 100-MeV nucleon is about 4×10^{-13} cm, and the kinetic energy transfer to the struck particle is about 25 MeV. Since the mean free path is comparable to nuclear radii, one cannot describe what goes on in terms of formation of a compound nucleus. In fact, what happens will depend on the particular trajectory of the incident particle. If it happens to pass through the nucleus near its edge, it may make a single collision and emerge having lost only 25 MeV of its energy, possibly having changed from neutron to proton (or vice versa) as a result of an exchange collision. Or, if it strikes the center of the nucleus and has to pass through the full diameter, it may make several collisions, lose all its energy, and end its range still inside the nucleus. There are thus a variety of possibilities, ranging from the bombarding particle emerging with most of its energy intact to the loss of the entire incident energy to the nucleus.

Since the struck particles have much lower energy and shorter mean free path than the incident one, they can escape from the nucleus without further collisions only if the collision occurs near the edge of the nucleus with the struck particle heading outwards. In this case it may emerge with 15- or 20-MeV energy. Otherwise it will collide with other nuclear particles, the energy will be distributed over the nucleus, and the subsequent events can be described in terms of the usual evaporation model, the nuclear excitation energy being dissipated by successive boiling off of particles each with a few million volts of kinetic energy.

When the nucleus is bombarded with 200-MeV deuterons, or 400-MeV α-particles, the binding of the incident nucleons is important chiefly in causing a spatial correlation between them, and what goes on can be thought of in terms of a simultaneous bombardment by several individual nucleons.

The description we have given provides an explanation of several features of high-energy reactions which have been observed at the Radiation Laboratory. Because of the wide distribution of excitation energies of the struck nucleus, one would expect a wide distribution of residual nuclei after the evaporation processes are complete; loss of a small number of particles should occur, as well as knocking out of many. This feature of high-energy reactions has been reported by Seaborg, Perlman, and their collaborators.[1] Then we may ask about the excitation function of a particular reaction leading to a given residual nucleus. At low energies the reaction proceeds through formation of a compound nucleus. If we confine ourselves to this mechanism, the excitation function will go through a maximum when the excitation energy is most appropriate for evaporating the requisite number of particles, then will drop very rapidly as the energy gets higher because at higher excitation energy it is much more likely that more particles will evaporate. However, at high energies the mechanism of the reaction is different, because of the transparency of the nucleus; the reaction can occur through the incident particle carrying off a good fraction of its energy and giving the nucleus approximately the right excitation energy for the reaction in question. Since the probability of leaving a given excitation energy will be determined only by the mean free path, which varies slowly with the energy of the incident particle, we would expect the excitation function at high energies also to vary quite slowly. This has been confirmed in a number of cases.[2]

Finally, we should expect the transparency of nuclear matter to show up in measurements of the total cross-section for absorption or scattering of the incident particle. It should be mentioned that the attenuation of the wave representing the incident particle in passing through the nucleus will give rise to diffraction scattering at small angles ($\theta \sim \hbar/Rp$, where R is the nuclear radius, p the momentum of the bombarding particle). The cross-section for diffraction scattering is equal to the inelastic and absorption cross-section; for good geometry attenuation measurements it just doubles the cross-section. For 100-MeV neutrons, with a mean free path of 4×10^{-13} cm, one sees that for the heaviest elements one would expect a total cross-section still close to $2\pi R^2$, but for light elements the cross-section should drop considerably below this value. This is, in fact, true, as has been shown by experiments by Cork, McMillan, Peterson, and Sewell.

A number of more detailed calculations, based on the considerations given above, have been carried out by members of the theoretical group at this laboratory. Reasonably good agreement has been obtained with experimental results on the excitation curves and absolute cross-sections of a few light element reactions, and on curves of star size *versus* frequency which have been measured by E. Gardner.[3] More detailed reports on this work will be published in the near future.

This paper is based on work performed under Contract W-7405-Eng-48 with the Atomic Energy Commission in connection with the Radiation Laboratory, University of California, Berkeley, California.

Notes

[1] B. B. Cunningham, H. H. Hopkins, M. Lindner, D. R. Miller, P. R. O'Connor, I. Perlman, G. T. Seaborg, and R. C. Thompson, *Phys. Rev.* **72**, 739 (1947).

[2] W. Chupp and E. M. McMillan; R. Thornton and R. W. Senseman, to be published.

[3] Eugene Gardner, *Phys. Rev.* **72**, 743 (1947).

10

Angular Distribution in (*d*, *p*) and (*d*, *n*) Reactions†

A. B. BHATIA, KUN HUANG, R. HUBY and H. C. NEWNS

Abstract. The angular distribution of the particles emitted in (*d*, *p*) or (*d*, *n*) reactions is calculated, assuming a stripping process, by means of the Born approximation. The result resembles in most respects that of more complicated calculations by S. T. Butler, but there are certain differences which are discussed in relation to experiments which have been, or may be done. The particles are emitted mainly at fairly small angles to the incident direction, and the precise shape of the distribution is simply related to changes of angular momentum and parity occurring. The principal unknowns in the formulae are matrix elements, which may be estimated for sufficiently high energies by the statistical method.

1. Introduction

Many measurements of the angular distribution in (*d*, *n*) and (*d*, *p*) reactions have been made recently using deuterons of energy in the region 8–15 MeV (e.g. Holt and Young, 1950; Burrows *et al.*, 1950; Hughes, 1951; El Bedewi *et al.*, 1951 a, b; Gove, 1951; Allen *et al.*, 1951).

The most pronounced feature in general is the existence of one or more peaks at fairly small angles to the incident direction. A simple theory to account for these results was developed by K. Huang and A. B. Bhatia, and briefly reported on by R. Huby (1950). This theory is based on the stripping mechanism as proposed by Serber (1947) to account for (*d*, *p*) and (*d*, *n*) results (including the angular distribution) at much higher energies (~ 200 MeV deuterons). At the lower deuteron energy now under con-

† *Phil. Mag.* **43**, 485 (1952).

sideration a simple Born approximation is applied to the stripping model, and a generally satisfactory form of angular distribution results. For outmoving neutrons or protons of definite energy, the form of the angular distribution is related in a simple way to the change of angular momentum and parity produced in the target nucleus by the capture of a nucleon from the incident deuteron.

The use of the Born approximation for a nuclear process at the energies concerned is difficult to justify theoretically; though as the form of the angular distribution turns out to be determined largely by such general features as changes of angular momentum and parity, it may be hoped that the results are largely independent of the details of treatment, rather in the same way as the dispersion formula for nuclear reactions can be obtained by a simple second-order perturbation treatment (e.g. Bethe, 1937) as well as by more correct methods.

Butler (1950, 1951) has also given a theory for the angular distribution in (d, p) and (d, n) reactions, based on the stripping mechanism. The mathematical technique used is not a perturbation one, but of a more penetrating kind involving the smooth joining up of wave functions for the captured nucleon at a radius r_0, and thus allowing rigorously for the strong interaction of this nucleon with the nucleus. However, the theory appears still to rely on some approximations whose effect is difficult to assess, notably the assumption that there is a region near the nucleus (radius $\sim r_0$) where one may neglect the interactions of the captured nucleon both with its former partner in the deuteron and with the target nucleus. It is believed that the development given in section 3 of Butler's (1951) paper contains some errors, though his final formula (34) is correct. For these reasons, and because of the complexity of the calculations of Butler, it is thought that the much simpler calculations of the Born approximation may be of interest, demonstrating as they do very clearly actually what happens in the reaction.

The formula obtained with the Born approximation (18, below) is very similar to Butler's in its general features. It contains a radial parameter R which is more arbitrary to fit correctly than the corresponding one (r_0) in Butler's formula, but otherwise it usually gives equally good agreement with experiment.

The angular distributions obtained with the two formulae differ most when the final nucleus is left in a state only just stable, or slightly unstable, against nucleon decay. The preliminary indications are that in such a case the Born formula may give the better agreement with experiment, but further experimental evidence is needed (see section 6). It should be noted that neither formula takes account of the Coulomb repulsion between the deuteron and the nucleus, so that it is necessary that the deuteron energy be sufficiently above the Coulomb barrier. Coulomb effects will presumably persist up to higher energies in the (d, n) than in the (d, p) reaction, because in the former the proton bearing the charge has to penetrate right to the nucleus, whereas in the latter the reaction can take place without the proton's ever coming so close (cf. Oppenheimer and Phillips, 1935).

The formula derived below (18) has been applied by H. C. Newns to experiments by El Bedewi *et al.* (1951 a, b) and El Bedewi (1952).

The apparent success of the Born approximation for calculating angular distributions in the stripping reaction suggests its application also to other reactions which may depend on a related mechanism. This has already been tried for the inelastic scattering of deuterons, and results obtained consistent with experiment (Huby and Newns, 1951).

The formula of Huang and Bhatia (18) is derived in section 2, discussed in section 3 and compared with that of Butler in section 4. In section 5 it is shown how the formula can be used to obtain information about the state of the final nucleus. An example is discussed in section 6. Appendix 1 deals with results obtainable by recourse to the statistical theory of nuclei.

2. Derivation of Formula for the Cross-section

The discussions following will refer specifically to the (d, p) reaction, but the (d, n) reaction is embraced by interchanging the roles of neutron and proton throughout.

ANGULAR DISTRIBUTION 279

In the stripping theory of the (d, p) reaction, it is assumed that the proton in the deuteron never comes into contact with the target nucleus: the neutron in the deuteron strikes the nucleus and is absorbed, while the proton continues its initial motion undisturbed, and emerges. We therefore only take into account an interaction of the neutron with the nucleus. The Born approximation for the differential cross-section is then given by:

$$\sigma = \frac{1}{4\pi^2 \hbar^4} M_p^* M_d^* \frac{k_p}{k_d} \mid I \mid^2, \qquad (1)$$

where

$$I = \int \chi_f^* (\boldsymbol{r}_n, \sigma_n, \xi) \chi_p^* (\sigma_p) e^{-i\boldsymbol{k}_p \cdot \boldsymbol{r}_p'} V (\boldsymbol{r}_n, \sigma_n, \xi) \chi_i (\xi) \chi_d (\boldsymbol{r}_n - \boldsymbol{r}_p, \sigma_n, \sigma_p) e^{[i\boldsymbol{k}_d \cdot (\boldsymbol{r}_n + \boldsymbol{r}_p)]/2} d\boldsymbol{r}_n \, d\boldsymbol{r}_p \, d\sigma_n \, d\sigma_p \, d\xi. \qquad (2)$$

The symbols have the meaning:

$$M_p^* = \text{reduced mass of proton} = \frac{M_p M_f}{M_p + M_f},$$

$$M_d^* = \text{reduced mass of deuteron} = \frac{M_d M_i}{M_d + M_i},$$

where M_p, M_n, M_d, M_i, M_f, are the masses of proton, neutron, deuteron, initial nucleus and final nucleus respectively; $k_p = (M_p^* v_p)/\hbar$, $k_d = (M_d^* v_d)/\hbar$, where v_p is the velocity of the outgoing proton relative to the centre of gravity of the final nucleus, and v_d is the velocity of the incoming deuteron in the laboratory system; $\boldsymbol{r}_n$, $\boldsymbol{r}_p$ are the coordinates of the neutron and proton relative to the centre of gravity of the initial nucleus; $\boldsymbol{r}_p'$ are the coordinates of the proton relative to the centre of gravity of the final nucleus ($\boldsymbol{r}_p' = \boldsymbol{r}_p - (M_n/M_f)\boldsymbol{r}_n$); σ_n, σ_p are the spin coordinates of neutron and proton; ξ represents all the internal coordinates of the initial nucleus; χ_p, χ_d, χ_i, χ_f are internal wave functions for the proton, deuteron, initial and final nuclei respectively; $V(\boldsymbol{r}_n, \sigma_n, \xi)$ is the interaction operator between the neutron and the particles in the initial nucleus.

We assume that the process of capture of the neutron takes

place close to the surface of the initial nucleus, or, more precisely, in some narrow shell just outside the surface, where we can roughly imagine the attraction of the nucleus for the neutron to outweigh that of the proton in the deuteron. Let the mean radius of this shell be R, which ought to be approximately the radius of the nucleus plus the range of nuclear forces. Mathematically, this assumption implies that the major contribution to the integral (2) comes from a small range of r_n close to R, and so in the factors of the integrand which represent the incident wave, we may replace $\boldsymbol{r} = (r_n, \theta_n, \phi_n)$ by $\boldsymbol{R} = (R, \theta_n, \phi_n)$ and obtain

$$I = \int \chi_p^*(\sigma_p) e^{-i k_p \cdot r_p'} \chi_d(\boldsymbol{R} - \boldsymbol{r}_p, \sigma_n, \sigma_p) e^{[i k_d \cdot (\boldsymbol{R} + \boldsymbol{r}_p)]/2}$$
$$\{\int \chi_f^*(r_n, \sigma_n, \xi) V(r_n, \sigma_n, \xi) \chi_i(\xi) r_n^2 \, dr_n \, d\xi\} dr_p \, d\sigma_p \, d\sigma_n \, d\omega_n. \quad (3)$$

The expression in the brackets is a function of the angular coordinates and the spin of the neutron. It may thus be expanded as follows:

$$\int \chi_f^*(r_n, \sigma_n, \xi) V(r_n, \sigma_n, \xi) \chi_i(\xi) r_n^2 \, dr_n \, d\xi$$
$$= \sum_{l_n, m_n, \mu_n} \langle f | V | i, l_n, m_n, \mu_n \rangle Y_{l_n}^{m_n *}(\theta_n, \phi_n) \chi_{\mu_n}^*(\sigma_n), \quad (4)$$

where $Y_{l_n}^{m_n}$ is a spherical harmonic, and $\chi_{\mu_n}(\sigma_n)$ is a spin eigenfunction. The expansion coefficients are given by:

$$\langle f | V | i, l_n, m_n, \mu_n \rangle = \int \chi_f^*(r_n, \sigma_n, \xi) V(r_n, \sigma_n, \xi) \chi_i(\xi)$$
$$Y_{l_n}^{m_n}(\theta_n, \phi_n) \chi_{\mu_n}(\sigma_n) dr_n \, d\sigma_n \, d\xi. \quad (5)$$

Substitution of (4) in (3) yields

$$I = \sum_{l_n, m_n, \mu_n} \langle f | V | i, l_n, m_n, \mu_n \rangle \langle l_n, m_n, \mu_n, p | d \rangle, \quad (6)$$

where

$$\langle l_n, m_n, \mu_n, p | d \rangle = \int Y_{l_n}^{m_n *}(\theta_n, \phi_n) \chi_{\mu_n}^*(\sigma_n) \chi_p^*(\sigma_p) e^{-i k_p \cdot r_p'}$$
$$\times \chi_d(\boldsymbol{R} - \boldsymbol{r}_p, \sigma_n, \sigma_p) e^{[i k_d \cdot (\boldsymbol{R} + \boldsymbol{r}_p)]/2} dr_p \, d\sigma_p \, d\sigma_n \, d\omega_n. \quad (7)$$

Now the components of angular momentum of the various internal wave functions viz. μ_p, μ_d, μ_i, μ_f are experimentally unspecified, and so the quantity $|I|^2$ required in (1) is rather an average over the initial and sum over the final components, i.e., with the aid of (6) we modify $|I|^2$ to:

$$|I|^2 = \frac{1}{3(2j_i+1)} \sum_{\substack{\mu_p,\mu_d \\ \mu_i,\mu_f}} \left| \sum_{l_n,m_n,\mu_n} \langle \mu_f | V | \mu_i, l_n, m_n, \mu_n \rangle \right. $$
$$\left. \langle l_n, m_n, \mu_n, \mu_p | \mu_d \rangle \right|^2. \tag{8}$$

Here the insertion of symbols such as μ_p in the Dirac brackets $\langle | \rangle$ indicates that internal wave functions χ_{μ_p}, etc., with the specified angular momentum are to be used in the integrals (5) and (7).

(8) can be simplified if the d-state component of the deuteron's wave function is neglected, i.e. if χ_d is a product of a triplet spin factor and an s-state orbital factor:†

$$\chi_{\mu_d}(\boldsymbol{R} - \boldsymbol{r}_p, \sigma_n, \sigma_p) = \psi(|\boldsymbol{R} - \boldsymbol{r}_p|) S_{\mu_d}(\sigma_n, \sigma_p).$$

Then (7) becomes

$$\langle l_n, m_n, \mu_n, \mu_p | \mu_d \rangle = g_{l_n}{}^{m_n} \langle \mu_n, \mu_p | \mu_d \rangle, \tag{9}$$

where

$$g_{l_n}{}^{m_n} = \int Y_{l_n}{}^{m_n*}(\theta_n, \phi_n) e^{-i\boldsymbol{k}_p \cdot \boldsymbol{r}_p'} \psi(|\boldsymbol{R} - \boldsymbol{r}_p|) e^{[i\boldsymbol{k}_d(\boldsymbol{R}+\boldsymbol{r}_p)]/2} d\omega_n \, d\boldsymbol{r}_p, \tag{10}$$

and

$$\langle \mu_p, \mu_n | \mu_d \rangle = \int \chi_{\mu_p}^*(\sigma_p) \chi_{\mu_n}^*(\sigma_n) S_{\mu_d}(\sigma_n, \sigma_p) \, d\sigma_n \, d\sigma_p. \tag{11}$$

After substituting (9) in (8), we can easily perform the summation over μ_p and μ_d with the result:

$$|I|^2 = \frac{1}{2(2j_i+1)} \sum_{\mu_n,\mu_i,\mu_f} \left| \sum_{l_n,m_n} \langle \mu_f | V | \mu_i, l_n, m_n, \mu_n \rangle g_{l_n}{}^{m_n} \right|^2. \tag{12}$$

† If the d mixture were allowed for, the result would probably be more complicated, and the selection rules governing l_n (Section 3) would probably be less stringent. These effects are, however, expected to be of small magnitude.

The summation over μ_n, μ_i and μ_f can now be carried out with the aid of the group-theoretical properties of the wave functions concerned in the definition of the Dirac bracket expressions. The result is

$$|I|^2 = \frac{1}{2(2j_i + 1)} \sum_{l_n, m_n} \frac{\Lambda_{l_n}}{(2l_n + 1)} |g_{l_n}{}^{m_n}|^2, \qquad (13)$$

where

$$\Lambda_{l_n} = \sum_{\substack{m'_n, \mu_n \\ \mu_i, \mu_f}} \left| \langle \mu_f | V | \mu_i, l_n, m'_n, \mu_n \rangle \right|^2. \qquad (14)$$

$g_{l_n}{}^{m_n}$ can readily be evaluated from (10). If the principal axis for determining l_n is taken in the direction of

$$\boldsymbol{k} = \boldsymbol{k}_d - \frac{M_i}{M_f} \boldsymbol{k}_p, \qquad (15)$$

$\left(\text{note } k = \left[\left(k_d - \frac{M_i}{M_f} k_p \right)^2 + 4 \frac{M_i}{M_f} k_p k_d \sin^2 \tfrac{1}{2}\theta \right]^{1/2}, \text{ where } \theta \text{ is the angle between } \boldsymbol{k}_p \text{ and } \boldsymbol{k}_d \right)$, then we find:

$$g_{l_n}{}^{m_n} = \delta_{m_n,0} [4\pi(2l_n + 1)]^{1/2} i^{l_n} G(|\boldsymbol{k}_p - \tfrac{1}{2}\boldsymbol{k}_d|) \left(\frac{\pi}{2kR} \right)^{1/2} J_{l_n + \frac{1}{2}}(kR), \qquad (16)$$

where

$$G(K) = \frac{4\pi}{K} \int_0^\infty \psi(r) r \sin Kr \, dr. \qquad (17)$$

Substitution of (16) into (13) and then into (1) yields

$$\sigma = \frac{M_p^* M_d^*}{2\pi(2j_i + 1)\hbar^4} \frac{k_p}{k_d} G^2(|\boldsymbol{k}_p - \tfrac{1}{2}\boldsymbol{k}_d|) \sum_{l_n} \Lambda_{l_n} \left[\left(\frac{\pi}{2kR} \right)^{1/2} J_{l_n + \frac{1}{2}}(kR) \right]^2. \qquad (18)$$

A formula for G is obtained if we use for the deuteron spatial wave function the approximate form:

$$\psi(r) = \left(\frac{\alpha}{2\pi}\right)^{1/2} \frac{e^{-\alpha r}}{r}, \qquad (19)$$

where $\alpha = 1/\hbar \times (M_p \times \text{deuteron binding energy})^{1/2}$

$$= 0 \cdot 23 \times 10^{13} \text{ cm}^{-1}. \qquad (20)$$

Then

$$G(|\,\boldsymbol{k}_p - \tfrac{1}{2}\boldsymbol{k}_d\,|)$$
$$= \frac{2(2\pi\alpha)^{1/2}}{\alpha^2 + |\,\boldsymbol{k}_p - \tfrac{1}{2}\boldsymbol{k}_d\,|^2} = \frac{2(2\pi\alpha)^{1/2}}{\alpha^2 + (k_p - \tfrac{1}{2}k_d)^2 + 2k_p k_d \sin^2 \tfrac{1}{2}\theta}. \qquad (21)$$

3. Interpretation of Results

The cross-section σ (18) contains a sum of terms over a parameter l_n which may be interpreted, in view of definitions (5) and (7), as the orbital angular momentum carried by the absorbed neutron into the nucleus. There are three important factors in the formula (18):

(i) Deuteron factor G^2 (17 and 21)

The proton, which initially has the mean momentum $\hbar k_d/2$ in the centre of mass frame, emerges with the momentum $\hbar k_p$. The difference $\hbar(\boldsymbol{k}_p - \boldsymbol{k}_d/2)$ represents the momentum of the proton in its motion inside the deuteron, at the instant of its release due to the removal of the neutron; and the factor $G^2(|\boldsymbol{k}_p - \boldsymbol{k}_d/2|)$ is proportional to the probability of this internal deuteron momentum. Equation (21) shows that this factor, as function of the angle θ between $\boldsymbol{k}_p$ and $\boldsymbol{k}_d$, has a forward maximum. The reason is that the larger the angle the larger the momentum difference which must be provided by the internal motion of the deuteron, and hence the smaller the probability. The factor G^2 applies equally to all transitions with different l_n.

(ii) Neutron centrifugal factor $[(\pi/2kR)^{1/2}J_{l_n+\frac{1}{2}}(kR)]^2$

From the last process, the neutron can be considered as issuing virtually from the deuteron with momentum $\hbar(\mathbf{k}_d - \mathbf{k}_p)$ in the centre of mass frame. After its capture by the target nucleus, the neutron has the mean momentum $-(M_n/M_f)\hbar\mathbf{k}_p$. The mean momentum interchanged between neutron and initial nucleus is thus $\hbar\mathbf{k} = \hbar[\mathbf{k}_d - (M_i/M_f)\mathbf{k}_p]$. The factor $[(\pi/2kR)^{\frac{1}{2}}J_{l_n+\frac{1}{2}}(kR)]^2$ is proportional to the mean probability that the neutron, travelling with this momentum in the frame in which the centre of gravity of the final nucleus is at rest, be found at the surface R of the nucleus, in a state in which the neutron's total orbital angular momentum is l_n.

With the aid of (15), we see that the present spherical Bessel function factor is an oscillatory function of the angle θ, the oscillations decreasing as θ increases. Provided $|\mathbf{k}_d - (M_i/M_f)\mathbf{k}_p|R$ is not too large, the factor has a maximum at the origin for $l_n = 0$, and for all other l_n a minimum at the origin, followed by a primary maximum which decreases in magnitude and moves to larger angles as l_n increases. The reason is that the larger the angular momentum l_n, the greater is the momentum $\hbar k$ (and hence the angle θ) which is necessary for the neutron to penetrate to the radius R.

(iii) Nuclear matrix element Λ_{l_n} (5 and 14)

This is proportional to the probability of capture of a neutron which is at the surface of the nucleus with initial orbital angular momentum l_n. We may note that Λ_{l_n} is determined only by the properties of the initial and final nuclear levels, and by l_n. When j_i and j_f and the initial and final parities are given, (14) shows clearly that there will be selection rules for the values of l_n which yield non-zero values of Λ_{l_n}. It follows from conventional group theoretical arguments that the selection rules are:

(a) The final angular momentum j_f must be obtainable by vector addition of the initial j_i, the orbital angular momentum l_n of the absorbed neutron, and the latter's spin angular momentum $1/2$.

(b) l_n must be even or odd, according as the parities of the initial and final levels are the same or different. These selection rules severely limit the number of terms appearing in (18), and can often bring it down to only one.

It is not easy to make predictions about the magnitudes, absolute or relative, of the Λ_{l_n}'s. Generally, it will be expected that different final states of similar constitution will have Λ_{l_n}'s of roughly the same order of magnitude. For example, for any final state in which the captured neutron is approximately moving in an orbit outside the initial nucleus in its original state, there will be only one value of l_n contributing appreciably, equal to the neutron's orbital angular momentum, even though more l_n's might be permitted by the selection rules (a) and (b). The Λ_{l_n}'s for such final states might be considerably larger than those for most others; and the relative values of the Λ_{l_n}'s for different final states of this type would be largely governed by a statistical factor $(2j_f + 1)$.

A rough estimate of the Λ_{l_n}'s may be made by the statistical method, as used by Bethe (1938) for calculations of the cross-section integrated over all angles. It is shown in the appendix that, for highly excited final states unstable against neutron emission, this yields

$$\Lambda_{l_n} = \frac{\hbar^2 k_n R^2 (2j_f + 1)\mu}{4\pi M_n^* \rho_{jf}(E_f)}. \tag{22}$$

Here k_n is the wave number of the neutron emitted if the final nucleus decays to the initial, i.e.

$$\frac{\hbar^2 k_n^2}{2M_n^*} = -\ (Q \text{ of } (d, p) \text{ reaction leading to final state } f$$
$$+ \text{ binding energy of deuteron}). \tag{23}$$

$\rho_{jf}(E_f)$ is the density of levels of the final nucleus in the neighbourhood of the actual final level; counting only levels of the same angular momentum j_f and parity as the level under consideration.

286 NUCLEAR REACTIONS

μ is 1 or 2, according as only one, or both, of the vectors of lengths $(j_i + \tfrac{1}{2})$ and $(j_i - \tfrac{1}{2})$ can be added vectorially to l_n to give a vector of length j_f (otherwise, of course, $\mu = 0$).

The derivation of (22) only applies for $k_n R > l_n$, but we may perhaps cautiously extrapolate it to smaller k_n, and even to imaginary k_n (corresponding to a state of the final nucleus stable against neutron emission), if we replace a factor $k_n R$ in the numerator by unity—for an order of magnitude result.

It is also shown in the appendix how the use of (22) in the cross-section (18), for very high deuteron energy, reproduces essentially some of the results on angular distribution and energy spectrum obtained by Serber (1947), who did not analyse in detail the changes occurring in the target nucleus. For example, summation of the differential cross-section over many neighbouring final levels yields a result in which the combined effect of the centrifugal factors (spherical Bessel functions) is independent of angle, and the residual angular distribution is due to the deuteron factor alone.

4. Comparison with the Formula of Butler (1951)

For greatest ease of comparison of formula (18) with the formula (34) of Butler (1951), the latter may be rearranged somewhat, and the notation modified, to yield:

$$\sigma = \frac{M_p^* M_d^*}{2\pi(2j_i + 1)} \frac{k_p}{\hbar^4 k_d} G'^2(|\,\boldsymbol{k}_p - \tfrac{1}{2}\boldsymbol{k}_d\,|)$$

$$\times \sum_{l_n} N_{l_n} e^{\kappa_n r_0} \left[\kappa_n r_0 K_{l_n - \tfrac{1}{2}}(\kappa_n r_0) \left(\frac{\pi}{2kr_0}\right)^{1/2} . J_{l_n + \tfrac{1}{2}}(kr_0) \right.$$

$$\left. + K_{l_n + \tfrac{1}{2}}(\kappa_n r_0) \left(\frac{\pi k r_0}{2}\right)^{1/2} J_{l_n - \tfrac{1}{2}}(kr_0) \right]^2. \quad (24)$$

Here G' is a function (deuteron factor) similar to G in (21), but differing a little because Butler used a more accurate form of wave function for the deuteron ground state than (19). r_0 is a nuclear radius. κ_n is defined by

$$\kappa_n = -ik_n, \tag{25}$$

(cf. eq. (23)), and is thus real if the final nucleus is stable against neutron decay. The case that the final nucleus is formed in a state unstable against neutron decay, and hence κ is imaginary, does not have to be treated separately in the Born calculations but it must be in Butler's. While Butler does not discuss this case fully, it is understood that the cross-section is still given by (24) provided the actual imaginary value of κ_n is inserted, and the square bracket is replaced by a modulus.

Remembering that the only variables containing the angle in (24) are k (eq. 15) and $|k_p - \frac{1}{2}k_d|$, we see that the only essential difference in the angular distribution between (24) and (18) lies in adding to the spherical Bessel function $(\pi/2kr_0)^{\frac{1}{2}}J_{l_n+\frac{1}{2}}(kr_0)$ another term proportional to $(\pi kr_0/2)^{\frac{1}{2}}J_{l_n-\frac{1}{2}}(kr_0)$. The effect of this is usually rather small, and either formula can usually be made to fit an experimental curve equally well, if suitable values of r_0 and R are chosen, the latter always having to be rather larger than the former.

The difference depends mainly on κ_n. It is very small for $|\kappa_n|$ large (state of final nucleus very stable or very unstable), and greatest for $|\kappa_n|$ small (state of final nucleus only just stable or just unstable), because in this case the second term in the square bracket in (24) takes control. The effects ensuing will be illustrated in an example in section 6. The conclusion is that Butler's formula behaves in a 'special' or 'singular' manner when $\kappa_n = 0$, while the Born formula does not, and the precise behaviour of the former will depend rather sensitively on the value of κ_n when small. It does not seem clear whether this singular behaviour is to be expected physically, and it would thus be of interest to make careful comparisons with experiment in this region.

5. Application to Determination of States

It is to be expected that the formula (18) will only agree with experimental curves in the region of small angles where the cross-section is large; since the constructive interference of many partial waves of definite orbital angular momentum for the proton and deuteron, which is really responsible for the maximum cross-section, should be more reliable than the details of the destructive interference which makes the cross-section small at large angles; and, in addition, there will probably be a background of neutrons produced by the competing process involving formation and decay of a compound nucleus, and these will tend to mask the stripping cross-section where it is small.†

To compare (18) with an experimental curve, it is necessary to know which A_{l_n}'s are present, and their value. For certain initial and final states more than one l_n may be permitted by the selection rules of section 3 (iii)—though the values permitted must be either all even or all odd. However, in practice, it has been found possible in most cases to obtain a satisfactory fit with one l_n only. When this has been found, if the parity and spin of the initial state are known then the selection rules fix the parity of the final level uniquely, and its spin j_f to within the limits permitted by vector addition of j_i plus l_n plus $\frac{1}{2}$.

To find l_n, the experimental curve should first be divided by the theoretical deuteron factor G^2 (see 21), so that the result is to be proportional simply to the sum in (18). A first indication of l_n can be obtained by inspection of this curve at small angles: a forward maximum will normally indicate that $l_n = 0$ is present, and a forward minimum that it is not (though these rules are not absolute). We can calculate the several spherical Bessel function terms in (18) and find which gives the best fit with experiment, if we take

† Cohen and Falk (1951) put forward experimental evidence from the energy spectrum in (d, n) reactions that the forward scattering is mainly due to stripping and that at 90° to compound nucleus. It is usually assumed that the angular distribution of particles from the compound nucleus will be fairly isotropic, but Wolfenstein's theory (1951) predicts peaking in the forward and backward directions, with symmetry about 90°. Indeed, it seems that Wolfenstein's theory could lead in extreme cases to a forward peak of sharpness comparable to that due on the stripping theory to the 'centrifugal' factor, section 3 (ii). However, it should be possible to tell whether such a forward peak is due to the stripping or the compound nucleus process, because the latter should be accompanied by a backward peak, and the former not.

a suitable value of R—and this last point, to choose the right R, is a major consideration. Butler (1951) found that with his formula (24) he usually obtained good and consistent agreement using for his radius r_0 an accepted value for the radius of the nucleus plus range of nuclear forces e.g. as given in Gamow and Critchfield (1949, p. 11):

$$\text{radius} = (1 \cdot 7 + 1 \cdot 22 \times A^{1/3}) \times 10^{-13} \text{ cm}; \qquad (26)$$

and on the average for light nuclei he took $5 \cdot 0 \times 10^{-13}$ cm. However, to obtain good and consistent results with the Born formula, a larger value of R is found to be needed, on the average $2 \cdot 0 \times 10^{-13}$ cm greater than the Gamow value (26).

The procedure is thus to find what value of l_n, together with a value of R not very different from that just mentioned, will give the best agreement with the experimental curve. The answer has in practice almost always turned out to be unique, for a given curve.

The large value of R required is hard to justify physically, and is in contrast to the reasonable value of the r_0 required in Butler's formula. Some investigation of the role of R or r_0 has therefore been made.

One hopes that the dependence of the angular distribution on R or r_0 will be insensitive, so that their exact choice is not very important. To obtain a measure of the sensitivity, we follow up a suggestion of Professor R. E. Peierls. Take the cross-section for $l_n = 0$ according to (18) and let k_0 be the value of k at which its first zero occurs. Then we may use as the gauge of sensitivity the quantity $(R/k_0)(dk_0/dR)$. Since in fact $k_0 \propto R$, this is just 1. The corresponding quantity $(r_0/k_0)(dk_0/dr_0)$ for Butler's formula (24) lies always between $0 \cdot 78$ and 1. This shows that the sensitivities of the two formulae are nearly the same.

Next we may analyse the statistics of a number of reactions for which both (18) and (24) have been fitted to the experimental curves. In each case the best values of R and r_0 to give a fit have been found, and compared to the Gamow radius (26). The results for 10 curves are that:

$$\left.\begin{array}{l}\text{for Born formula (18), } R = \text{Gamow radius} \\ \quad + 2\cdot 0 \times 10^{-13} \pm 0\cdot 50, \\ \text{for Butler's formula (24), } r_0 = \text{Gamow radius} \\ \quad + 0\cdot 36 \times 10^{-13} \pm 0\cdot 44.\end{array}\right\} \quad (27)$$

(The limits are 'probable errors' = $0\cdot 67 \times$ r.m.s. deviation from mean.) Thus the spread of the best values of the radius is about the same for both the Born and Butler's theories; but on the average Butler's radius comes out closer to the physically reasonable Gamow radius. Both methods appear in practice to be about equally reliable for finding the characteristics of the final state.

6. Example

We shall discuss only the results of the reactions $O^{16}(d, n)F^{17}$ and $O^{16}(d, n)F^{17*}$, which have been obtained by El Bedewi et al. (1951 a) using 8 MeV deuterons. The Q's of the respective reactions are $- 1\cdot 36$ MeV and $- 1\cdot 89$ MeV, yielding, in accordance with (25) and (23) the respective values for κ_p† of $0\cdot 196 \times 10^{13}$ cm^{-1} and $0\cdot 119 \times 10^{13}$cm^{-1} the latter in particular being unusually small.

Considering first the reaction forming F^{17} in the ground state, we find that both eqs. (18) and (24) require l_p† $= 2$ for a good fit. The radius R giving the best fit by Born approximation is $6\cdot 25 \times 10^{-13}$cm (note that the Gamow radius from (26) is $4\cdot 8 \times 10^{-13}$ cm, and compare with (27)); while with Butler's formula, a good result is obtained with the radius $5\cdot 0 \times 10^{-13}$ cm which he recommends on the average. The results are shown in Fig. 1. The insensitivity of the results to the value of the radius is shown by the curves of Fig. 2, which are drawn for radii r_0 and R each increased by 1×10^{-13} cm from the values mentioned above: the fit is still quite good. The reliability of the value of l_p is indicated by the fact that, taking the Born approximation, for instance, to obtain a good fit assuming $l_p = 1$ or 3, we should require respective values of R

† Note that, as this is a (d, n) reaction, the roles of neutron and proton are interchanged from those in the preceding discussions.

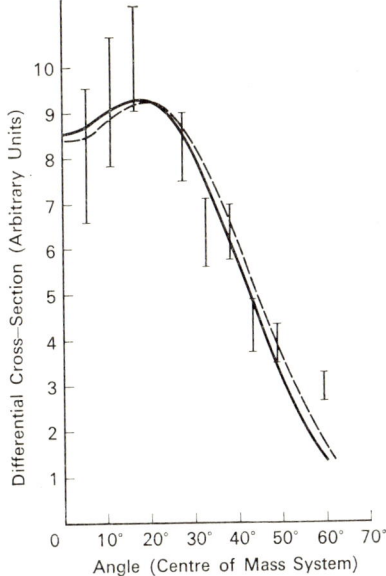

Angle (Centre of Mass System).

FIG. 1. Reaction $O^{16}(d, n) F^{17}$ with 8 MeV deuterons. $Q = -1.36$ MeV. Points are measured by El Bedewi et al. (1951a). Full curve: theory according to Born approximation with $R = 6.25 \times 10^{-13}$ cm. Broken curve: theory according to Butler, with $r_0 = 5 \times 10^{-13}$ cm.

of 3.9×10^{-13} and 8.4×10^{-13} cm, neither of which conforms at all well to (27).

The conclusion to be drawn from the selection rules of section 3 (iii) is that, since the ground state of O^{16} has $j_i = 0$, even parity, and since $l_p = 2$, then the ground state of F^{17} has the spin $j_f = 3/2$ or $5/2$ and even parity.

For the first excited state, both formulae (18) and (24) require $l_p = 0$. The best values of R and r_0 for a fit are respectively 7.25×10^{-13} and 4.2×10^{-13} cm, and the curves for these are shown in Fig 3. Increasing r_0 to the 'average' value quoted by Butler, 5.0×10^{-13} cm, gives poor agreement, as shown in Fig. 4, whereas an increase of R in the formula (18) to 8.05×10^{-13} does not cause so

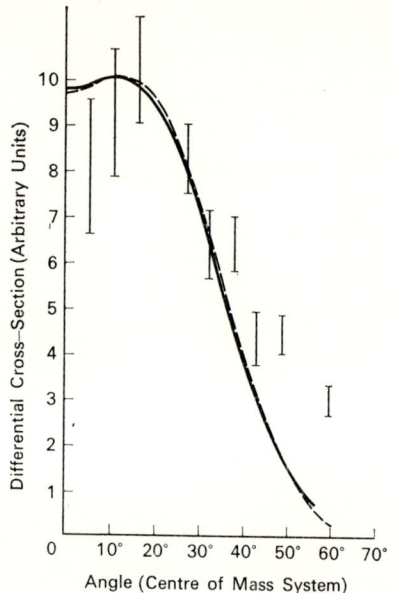

Fig. 2. Reaction O^{16} (d, n) F^{17} with 8 MeV deuterons. $Q = -1.36$ MeV. Points are measured by El Bedewi et al. (1951a). Full curve: theory according to Born approximation, with $R = 7.25 \times 10^{-13}$ cm. Broken curve: theory according to Butler, with $r_0 = 6 \times 10^{-13}$ cm.

much deterioration. The reason is the 'singular' behaviour of Butler's formula for small κ_p (see section 4), two of the chief features of which can be seen in this example, viz. (i) a reduction of the value of kr_0 at which the first zero of the cross-section occurs (tending to require unusually small values of r_0) and (ii) unusually large secondary maxima at large angles.†

† Although there is as yet not much evidence under these 'singular' conditions, it is of some interest that in one other case, C^{12} (d, n) N^{13*} (see El Bedewi et al., 1951 b) forming the final nucleus in its first excited state, which is slightly unstable against neutron decay ($\kappa_n = -i \times 0.14 \times 10^{13}$ cm^{-1}), it does not prove possible to obtain a really good fit for Butler's formula with any value of r_0: the nearest fit is got with $r_0 = 3.7 \times 10^{-13}$ cm, whereas $r_0 = 5 \times 10^{-13}$ cm is very poor.

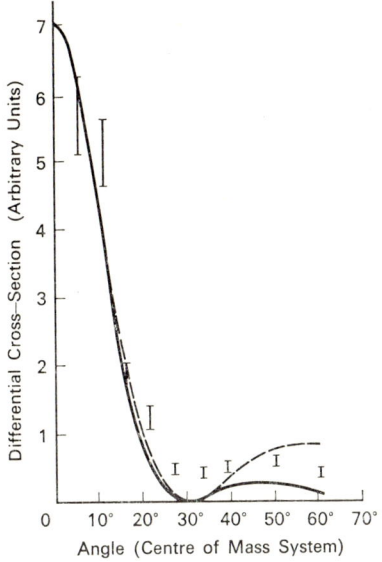

Angle (Centre of Mass System).

Fig. 3. Reaction $O^{16}(d, n) F^{17*}$ with 8 MeV deuterons. $Q = -1 \cdot 89$ MeV. Points are measured by El Bedewi *et al.* (1951a). Full curve: theory according to Born approximation with $R = 7 \cdot 25 \times 10^{-13}$ cm. Broken curve: theory according to Butler, with $r_0 = 4 \cdot 2 \times 10^{-13}$ cm.

The result $l_p = 0$, together with the known spin and parity of O^{16}, leads to the result $j_f = \frac{1}{2}$, and even parity, for the first excited state of F^{17}.

Appendix 1

USE OF THE STATISTICAL THEORY OF NUCLEI (FOLLOWING BETHE, 1938)

We shall express the matrix element Λ_{l_n} (14) in terms of the partial width for decay of the final nucleus with neutron emission, leaving the initial nucleus in its initial state (assuming this to be energetically possible). The partial width can then in turn be estimated in terms of a sticking probability.

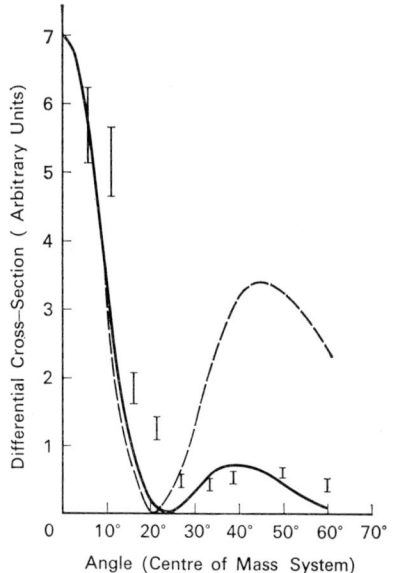

Fig. 4. Reaction $O^{16}(d, n) F^{17*}$ with 8 MeV deuterons. $Q = -1\cdot89$ MeV. Points are measured by El Bedewi et al. (1951 a). Full curve: theory according to Born approximation, with $R = 8\cdot05 \times 10^{-13}$ cm. Broken curve: theory according to Butler, with $r_0 = 5 \times 10^{-13}$ cm.

The formula for the partial width for the decay process in which the neutron is emitted with angular momentum l_n is, by perturbation theory:

$$\Gamma_{J_{i'},l_n;J_f} = \frac{4k_n}{\hbar^2(2j_f+1)} \sum_{\substack{\mu_i,\mu_f \\ \mu,n^m{}_n}} \left| \int \chi^*_{\mu i}(\xi) \chi^*_{\mu n}(\sigma_n) Y_{l_n}{}^{m_n*}(\theta_n,\phi_n)(\pi/2k_n r_n)^{1/2} J_{l_n+\frac{1}{2}}(k_n r_n) \right. $$
$$\left. \times V(r_n, \sigma_n, \xi) \chi_{\mu f}(r_n, \sigma_n, \xi) dr_n\, d\sigma_n\, d\xi \right|^2. \qquad (28)$$

(The notations are those of section 2; k_n was defined in (23).) If we make the assumption, as in section 2, that the interaction is effective only at the radius R, so that r_n in the spherical Bessel function (28) may be replaced by R, then with the help of (5) and (14) we obtain at once:

$$\Gamma_{J_{i'},l_n;J_f} = \frac{4k_n A_{l_n}}{\hbar^2(2j_f+1)} \left[\left(\frac{\pi}{2k_n R}\right)^{1/2} J_{l_n+\frac{1}{2}}(k_n R) \right]^2. \qquad (29)$$

ANGULAR DISTRIBUTION 295

If $k_n R$ is appreciably larger than $(l_n + \frac{1}{2})$, the asymptotic form of the Bessel function may be used, yielding on an average over a range of k_n:

$$\Gamma_{J_{l'},l_n;J_f} = \frac{2\Lambda_{l_n}}{\hbar^2(2j_f + 1)k_n R^2}. \tag{30}$$

Now by the principle of detailed balance the partial width is expressible in terms of the capture probability for a neutron of wave number k_n colliding with the initial nucleus (see, e.g. L. Wolfenstein, 1951)

$$\Gamma_{J_{l'},l_n;J_f} = \frac{(2l_n + 1)(2j_i + 1)\xi_{J_f;J_{\mu'},l_n}}{\pi(2j_f + 1)\rho_{J_f}(E_f)}. \tag{31}$$

$\rho_{J_f}(E_f)$ was defined below (23). $\xi_{J_f; J_i, l_n}$ is the sticking probability for the neutron, if its orbital angular momentum is l_n, to combine with the initial nucleus j_i to form a compound nucleus with the angular momentum j_f and proper parity.†

Now if we assume that the sticking probability does not depend explicitly on the final angular momentum j_f, but is determined only by statistical factors, we may write

$$\xi_{J_f; J_{l'},l_n} = \frac{(2j_f + 1)\mu\xi_{J_i, l_n}}{2(2l_n + 1)(2j_i + 1)}, \tag{32}$$

where μ was defined below (23), and ξ_{J_i, l_n} is the sticking probability irrespective of final state, which to the present accuracy may be taken as 1. Substituting (32) in (31) and thence in (30), we obtain the final formula (22) for Λ_{l_n}.

We now investigate the cross-section obtained in measuring many proton groups together. We find first the differential cross-section $d\sigma(\theta)$ for all protons in the energy range dE_p scattered at an angle θ, from (18):

$$d\sigma(\theta) = \frac{M_p^* M_d^*}{2\pi(2j_i + 1)\hbar^4} \cdot \frac{k_p}{k_d} G^2(|k_p - \tfrac{1}{2}k_d|) \cdot dE_p \sum_{l_n, J_f} \Lambda_{l_n} \left[\left(\frac{\pi}{2kR}\right)^{1/2} J_{l_n + \frac{1}{2}}(kR)\right]^2$$
$$\times \rho_{J_f}(E_f)$$

(substituting from (22) and summing over j_f)

$$= \frac{G^2(|k_p - \tfrac{1}{2}k_d|)R^2}{4\pi^2 \hbar^2} \frac{M_d k_p k_n}{k_d} dE_p \sum_{l_n} (2l + 1) \left[\left(\frac{\pi}{2kR}\right)^{1/2} J_{l_n + \frac{1}{2}}(kR)\right]^2.$$
$$= \frac{G^2(|k_p - \tfrac{1}{2}k_d|)R^2}{4\pi^2} \frac{M_d k_p k_n}{k_d} dE_p, \tag{33}$$

since the sum over l_n is just unity. (In the above we have put $M_n^* = M_n$ etc.) The only angular dependence here is through the deuteron factor G^2. (21) shows that, for high deuteron energy, this has a sharp peak, as function of

---

† The sticking probability as here used has to contain also any inhibiting effects of barrier penetration.

k_p, when $k_p = \frac{1}{2}k_d$. For important cases, then, we may put in (33)

$$k_p = k_n = \frac{k_d}{2},$$

and obtain

$$d\sigma(\theta) = G^2(|k_p - \tfrac{1}{2}k_d|)\frac{R^2 M_n k_d}{8\pi^2 \hbar^2} dE_p. \tag{34}$$

We can integrate (34) over all proton energies E_p, with the aid of (21); to obtain a total angular distribution $\sigma_{\text{tot}}(\theta)$ (after approximating for large k_d):

$$\sigma_{\text{tot}}(\theta) = \frac{R^2 k_d^2 \alpha}{(\alpha^2 + \tfrac{1}{4}k_d^2 \sin^2\theta)^{3/2}}. \tag{35}$$

The energy spectrum $d\sigma$ for all protons in the range dE_p is obtained similarly by integrating (33) over all angles θ with the aid of (21):

$$d\sigma = \frac{16 R^2 M_p \alpha k_p k_n dE_p}{\hbar^2 k_d[\alpha^2 + (k_p - \tfrac{1}{2}k_d)^2](k_p + \tfrac{1}{2}k_d)^2}. \tag{36}$$

The results (35) and (36) agree essentially with the eqs. (11) and (15) of Serber (1947) for the angular distribution and spectrum assuming a "transparent" nucleus, i.e. neglecting entirely the interaction of the proton with the target nucleus (only in Serber's paper it would be the neutron, not proton, since he discusses (d, n) reactions).

The total cross-section σ_{tot} for all angles and all proton energies is obtained by integrating either (35) over angles or (36) over E_p:

$$\sigma_{\text{tot}} = 2\pi R^2, \tag{37}$$

which is of the right order of magnitude, though too large.

References

ALLEN, A. J., NECHAJ, J. F., SUN, K. H., and JENNINGS, B. (1951) *Phys. Rev.* **81**, 536.
BETHE, H. A. (1937) *Rev. Mod. Phys.* **9**, 101; (1938) *Phys. Rev.*, **54**, 39.
BURROWS, H. B., GIBSON, W. M., and ROTBLAT, J., (1950) *Phys. Rev.*, **80**, 1095.
BUTLER, S. T. (1950) *Proceedings of the Harwell Nuclear Physics Conference* (Min. of Supply, Harwell), p. 54; (1951) *Proc. Roy. Soc.* **A208**, 559.
COHEN, B. L. and FALK, C. E. (1951) *Phys, Rev.* **84**, 173.
EL BEDEWI, F. A., MIDDLETON, R., and TAI, C. T. (1951a) *Proc. Phys. Soc.* **A64**, 756; (1951b) *Ibid.*, **64**, 1055.
EL BEDEWI, F. A. (1952) *Proc. Phys. Soc.* **A65**, 64.
GAMOW, G., and CRITCHFIELD, C. L. (1949) *Theory of Atomic Nucleus and, Nuclear Energy Sources* (Oxford, Clarendon Press).
GOVE, H. E. (1951) *Phys. Rev.*, **81**, 364.
HOLT, J. R., and YOUNG, C. T. (1950) *Proc. Phys. Soc.* **A63**, 833.

HUBY, R., (1950) *Nature, Lond.*, **166,** 552.
HUBY, R., and NEWNS, H. C. (1951) *Phil. Mag.*, **42,** 1442.
HUGHES, J. (1951) *Proc. Phys. Soc.* **A64,** 797.
OPPENHEIMER, J. R., and PHILLIPS, M. (1935) *Phys. Rev.*, **48,** 500.
SERBER, R. (1947) *Phys. Rev.*, **72,** 1008.
WOLFENSTEIN, L., (1951) *Phys. Rev.*, **82,** 690.

11

Elastic and Inelastic Diffraction Scattering†

J. S. BLAIR

I WOULD like to discuss briefly both the theory of inelastic diffraction scattering and some applications of this to inelastic scattering experiments performed primarily with strongly absorbed α-particles. Let us first consider some data which are typical for the elastic scattering of 40 MeV α-particles on light nuclei. The data are those of Yavin and Farwell. Figure 1 is intended to convince you that it is a relatively good approximation to consider the nucleus to be black to α-particles. The solid curve is the Fraunhofer black disk formula of physical optics. The one parameter, the radius, is chosen to match the periodicity and thus also the magnitude is completely determined, i.e. there is no arbitrary normalization. At small angles one has Coulomb complications—further, the simple expression fails for higher orders in the diffraction pattern because of the sharp edge assumption. But for low orders, the black disk formula appears relevant.

Next consider a typical comparison of elastic and inelastic scattering exciting a 2+ level, namely that for Mg^{24}. We note the following qualitative features of the figure: both curves show marked oscillations, there is a definite phase relationship between the two curves and the inelastic cross-section is moderately large compared to the elastic cross-section.

The time-honored manner of discussing inelastic scattering is to employ the plane wave Born approximation. However, we have just seen that the nucleus is far from transparent and thus

† *Proc. Int. Conf. Nuclear Structure*, Kingston, p. 824, ed. D. A. Bromley and E. W. Vogt, University of Toronto Press, Toronto, 1960.

distortion is very important. The large distortion can be included in distorted wave Born approximation computations and we have seen the very nice results obtained with this method by Rost and Austern.

However, it would be nice to have a simpler way of looking at the problem and, in particular, to understand why there is the close connection between elastic and inelastic scattering. Such a mechanism has been provided by the inelastic diffraction scattering model introduced by Drozdov and Inopin to treat scattering

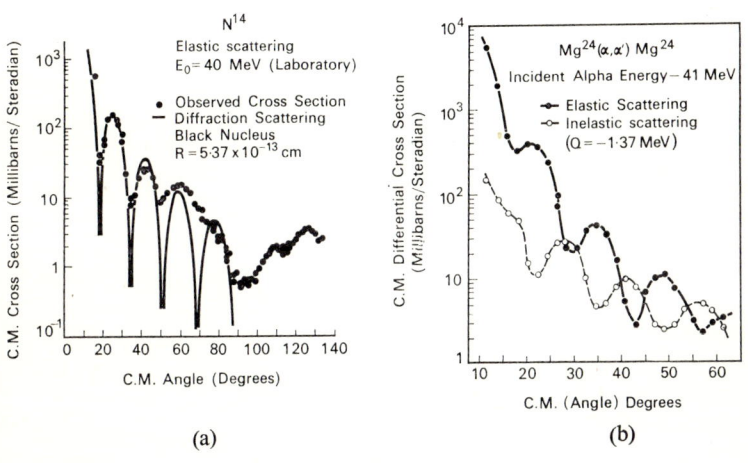

Fig. 1.

from a black ellipsoid. I will quickly review and generalize the theory.

The crucial step which allows us to relate elastic and inelastic scattering is the use of the adiabatic approximation. In general the scattering amplitude between two states a and b can be written in terms of the operator formalism as

$$<b\,|\,T\,|\,a> \,=\, <b\,|\,V + V\frac{1}{E - K - V - H(\alpha) + i\epsilon}\,V\,|\,a> \tag{1}$$

where the Hamiltonian for the system is

$$H = K + V(r, \alpha) + H(\alpha). \tag{2}$$

K is the kinetic energy of the projectile, $V(r, \alpha)$ is a potential which is a function of the coordinates of the projectile and a set of collective coordinates symbolized by α, while $H(\alpha)$ is the nuclear Hamiltonian. The states $|a>$ and $|b>$ are eigenstates of the reduced Hamiltonian,

$$K + H(\alpha),$$

i.e. $$|a> = e^{ik_a \cdot r} \mu_a(\alpha). \tag{3}$$

The adiabatic approximation is equivalent to freezing the motion of the nucleus during the time of the collision. Formally we represent that by striking $H(\alpha)$ in the general expression for a scattered amplitude. We also disregard the difference between the incident energy and final energy of the projectile. It is then a rigorous statement in this approximation that the scattered amplitude between the two states is equal to the elastic scattered amplitude for fixed collective coordinates sandwiched between the nuclear wave functions which are functions of these coordinates;

$$< b | T | a > \rightarrow < \mu_b(\alpha) | f(\alpha, \theta) | \mu_a(\alpha) >. \tag{4}$$

This is now the step which connects an elastic to an inelastic process.

Another way of looking at the problem is to say that, for the purposes of calculating the inelastic cross-section, we pretend that the nucleus has an infinite moment of inertia. Thus all of the nuclear states would be degenerate and the elastic scattered amplitude induces transitions from one state to another. Essentially the inelastic excitation is then a form of Raman scattering.

Our chore now in this approximation is to compute an expression for the scattering amplitude into a given angle for a fixed set of deformation parameters. One can make a series of terrifying assumptions in order to get some simple expressions for such a scattered amplitude. Let us assume that we have strong absorption within a sharp radius, R, which is connected in the typical way to

ELASTIC AND INELASTIC DIFFRACTION SCATTERING

the collective coordinates,

$$R = R_0(1 + \Sigma_{lm}\, a_{lm}\, Y_{lm}). \tag{5}$$

Let us use the Fraunhofer approximation in which a space dependent phase factor is integrated over the projection of the nuclear volume onto the plane perpendicular to the beam.

$$f(\alpha, \theta) \cong \frac{ik}{2\pi} \int\int dA\, e^{-ik_f \cdot r} \tag{6}$$

We will also approximate the scattered amplitude only to terms linear in the deformation parameters, a_{lm}.

Physically what these approximations all amount to is that we are cutting a non-circular hole in the incident beam instead of a circular hole. This then gives us a scattered amplitude which is a function of the collective coordinates and thus the scattered amplitude projects on to excited nuclear state wave functions.

The relevant equations used in the calculation are those for the scattered amplitude with fixed a_{lm} using the aforementioned approximations.

$$f(\alpha, \theta) \cong \frac{ik}{2\pi} \int_0^{2\pi} d\phi \left[\int_0^{R_0} \exp(-ikr\,\theta \cos\phi) r\, dr \right.$$
$$\left. + \exp(-ikR_0\,\theta \cos\phi) R_0^2\, \Sigma_{lm}\, a_{lm}\, Y_{lm}(\pi/2, \phi) \right] \tag{7}$$

$$= ikR_0^2 \left[\frac{J_1(kR_0\,\theta)}{(kR_0\,\theta)} + \Sigma_{l,\,m \atop l+m\,\mathrm{even}} \left| \frac{2l+1}{4\pi} \right|^{\frac{1}{2}} i^l \right.$$
$$\left. \times \frac{[(l-m)!\,(l+m)!]^{\frac{1}{2}}}{(l-m)!!(l+m)!!}\, a_{lm}\, J_{|m|}(kR_0\,\theta) \right] \tag{8}$$

The first term in eq. (7) is an integral from 0 to R_0, the equilibrium distance, and this term gives just the usual black disk formula. The second term in eq. (7) is the contribution from the rim; when we integrate over ϕ, we encounter the defining relations for Bessel functions.

We are then led to the following simple expressions for the cross-sections:

$$\frac{d\sigma_{el}}{d\Omega} = (kR_0^2)^2 \left| \frac{J_1(x)}{x} \right|^2 \qquad (9)$$

$$\frac{d\sigma}{d\Omega}(0 \to 0) = (kR_0^2)^2 \frac{<0|\alpha_0|0>^2}{4\pi} J_0^2(x) \qquad (10)$$

$$\frac{d\sigma}{d\Omega}(0 \to 2) = (kR_0^2)^2 \frac{\beta_2^2}{4\pi} \left\{ \frac{1}{4} J_0^2 + \frac{3}{4} J_2^2 \right\} \qquad (11)$$

$$\frac{d\sigma}{d\Omega}(0 \to 3) = (kR_0^2)^2 \frac{\beta_3^2}{4\pi} \left\{ \frac{3}{8} J_1^2 + \frac{5}{8} J_3^2 \right\} \qquad (12)$$

where we have listed the elastic scattering cross-section, and the inelastic cross-sections for monopole, quadrupole, and octupole excitation, respectively. Here k is the wave number, and x is the traditional phase factor which depends upon our choice of shadow plane; the canonical choices give $x = 2k R_0 \sin(\theta/2)$ or $k R_0 \theta$. The J_l are regular, not spherical, Bessel functions.

Finally, we see that the octupole formula involves Bessel functions of odd order while the monopole and quadrupole formulas contain Bessel functions of even order. This is a general characteristic of the formulas; odd multipoles have odd orders, even multipoles have even order. Thus, we are rather naturally led to the observed phase relationships between the elastic and inelastic angular distributions since the Bessel functions break very quickly into sinusoidal functions of x. Specifically

$$J_l^2(x) \to \frac{2}{\pi x} \sin^2 \left(x + \frac{\pi}{4} - \frac{l\pi}{2} \right). \qquad (13)$$

Therefore the octupole angular distribution is in phase with the elastic cross-section while the monopole and quadrupole cross-section are out of phase with the elastic pattern.

We have couched the quadrupole and octupole matrix elements in the language of permanent deformation parameters, β_l. What counts in these formulas, however, is the magnitude of the collective matrix elements irrespective of their detailed evaluation. They may also be characterized by the parameters relevant to harmonic vibrations, namely the phonon excitation energy, ω_l, and the surface tension, C_l. The connection between these two parametrizations is the familiar formula

$$\beta_l^2 = (2l + 1) \frac{\hbar \omega_l}{2 C_l} \tag{14}$$

We note several features of these formulae, if we divide all of the cross-sections by k^2 and then plot these quotients versus the parameter x, then we should have universal curves. Also we observe that only the quantity x enters into the elastic scattering; this means that we essentially get a radius from analyzing elastic scattering alone. Thus the magnitudes of the inelastic curves are then functions only of the collective matrix elements.

Let us then test this model by measuring cross-sections at several different energies and then seeing if the scaling law works out. This has been done by Farwell and McDaniels. Figure 2 shows what happens when they take all of the available elastic α-particle data for Mg and plot it in such a fashion. The dashed curve which goes through squares with crosses in them, corresponds to deuterons; all the other points, however, are for α-particles. At the first and third maxima of the α-particle data there is a definite clustering of the points. At small angles there are Coulomb complications and at the higher orders of the pattern there is considerable scatter. Clearly you can look at this and say "so what", because we already observed that the elastic scattering was given by a black disk formula.

Let us now, however, take a look at the inelastic scattering for which we would make the same scaling prediction (Fig. 3). Again the squares with crosses are the deuteron data which show less structure and which generally lie below the α work. All the other points refer to α-particle data and here we see that the data do pinch together at the maxima, although they scatter at large angles. The solid line here is not a theoretical line, but rather

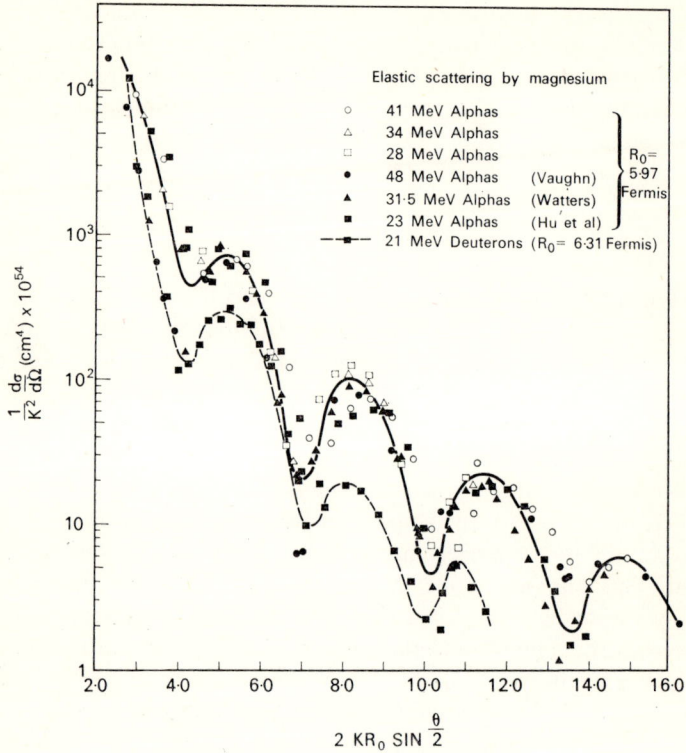

FIG. 2. Collected data on elastic α-particle scattering from magnesium together with 21 MeV deuteron data.

characterizes all of the assorted data which will now compare to the actual formulas in Fig. 4.

The top two curves refer to elastic scattering, the bottom two to inelastic scattering. The solid curve is the Fraunhofer formula, the dashed curve symbolizes the data. We show with crosses some points corresponding to the highest energy α-particle experiments where we expect the Coulomb effects to be minimized; we note that the points are closer to the black disk formula than is the dashed line. One notes that at the higher orders the experimental

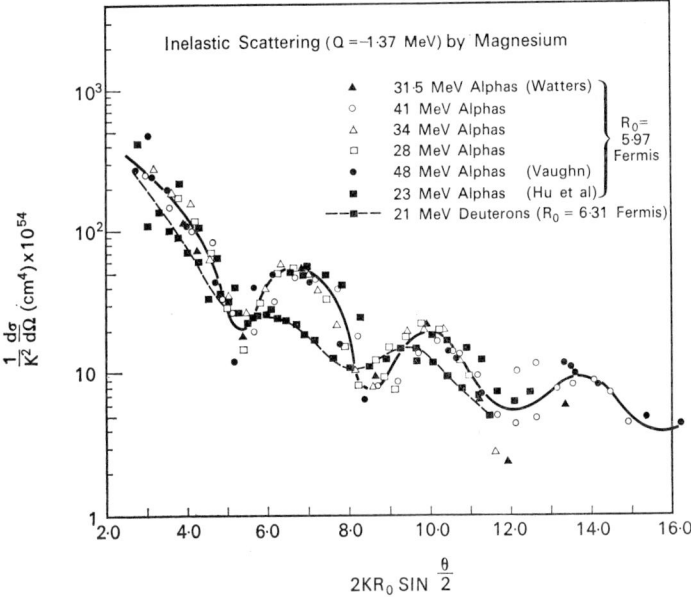

FIG. 3. Collected data on inelastic α-particle scattering involving the 1.37 MeV state in Mg^{24}; 21 MeV deuteron data is included for comparison.

results are very definitely under the sharp edge predictions. But in the region of the first two elastic maxima the simple black disk prediction seems to work. Now the rule of the game will be to normalize the inelastic cross-section only in the region where we have had some quantitative fit to the magnitude of the elastic cross-sections. Accordingly the inelastic curves will be matched at the point around $x \sim 7$. The theoretical inelastic curve shares the same deficiencies at small and large angles as does the elastic curve.

From such matching one can estimate the collective matrix elements. It is convenient for purposes of comparison to other experiments to consider not the parameter β_2 but rather the

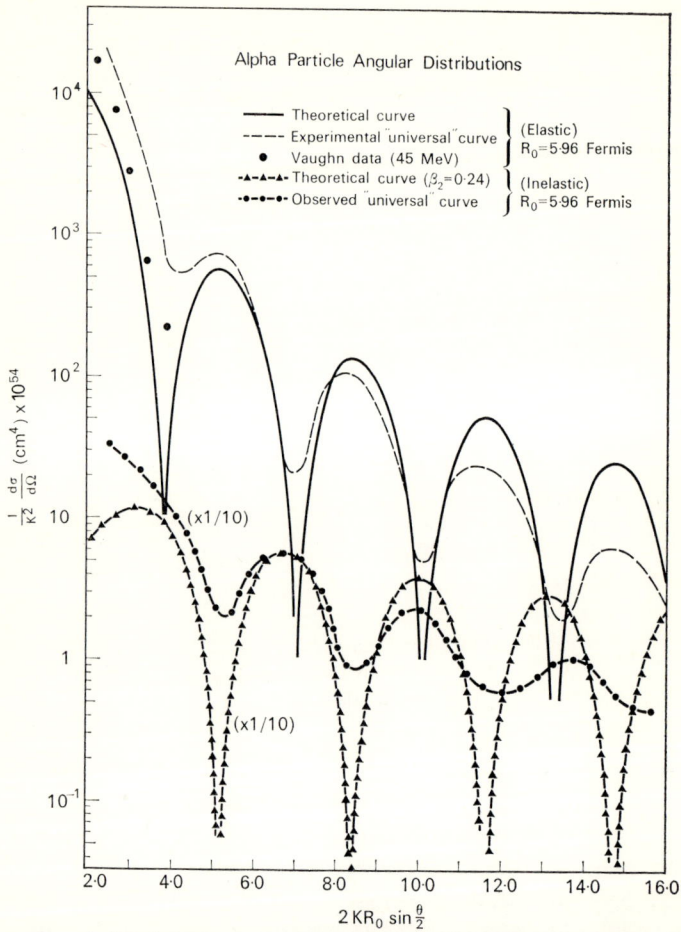

Fig. 4. "Universal" curves for elastic and inelastic α-particle scattering.

distance characteristic of the truly nuclear deformation, $\beta_2 R_0$. From the above normalization we obtain the value, $\beta_2 R_0 = 1.43$ fermis. How does this compare with other determinations? Electromagnetic measurements of course also determine this quantity, where we must use now the value of R which is appropriate to the mean radius of the electric charge distribution. We then obtain, for the electromagnetic estimates of this quantity, 1.7 or 2.3 fermis depending on two different choices of data.

Now let us ask the question, how does this compare with the exact distorted wave Born calculations? Figure 5 has previously been used by Austern. This illustrates that a distorted wave Born

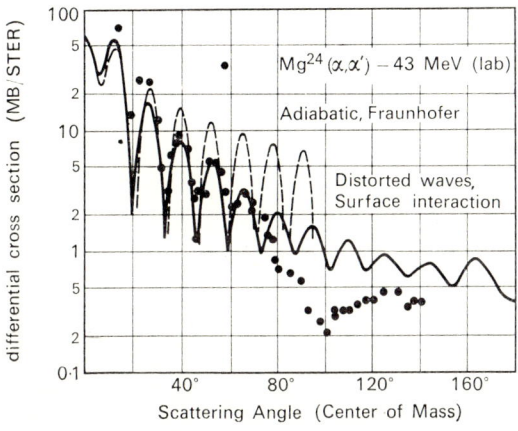

FIG. 5. Comparison of Fraunhofer and DWBA fits to 43 MeV inelastic α-particles scattering data.

calculation does follow the large angle behavior much better than does the Fraunhofer result. One should remark that this distorted wave calculation does not play around with changing the optical parameters; in this sense it too is a one parameter theory—the only free parameter is β_2. I should also say that in the region of smaller angles there is fair agreement between the Fraunhofer formula and the distorted wave calculation, but then, of course, the Fraunhofer oscillations wander off to a high value for larger angles. We can now ask, what is the value of $\beta_2 R_0$ obtained from this distorted wave calculation of Rost and Austern? We must compare in this fashion because the R_0 of these computations is a mean radius of some Saxon potential and is considerably less than the strong absorption radius of the Fraunhofer formula. The result is that $\beta_2 R_0 (\text{DWB}) = 1.67$.

Armed with such a comparison with experiment and to the DWB computation we are then tempted to go through the p and s-d shells and make a systematic comparison between α-particle inelastic cross-sections and this model for experiments in which the elastic cross-sections seem to be given by the black nucleus expression. We show in Table 1 some results for the case of quadrupole excitations.

TABLE 1. TABULATED VALUES OF THE NUCLEAR DEFORMATION $\beta_2 R_0$ OBTAINED FROM ANALYSIS OF α-PARTICLE SCATTERING EXPERIMENTS.

Nucleus	E_x MeV	J^π	$\beta_2 R_0$ fermis
Be^9	2.43	5/2−	2.3
C^{12}	4.43	2+	1.4
Ne^{20}	1.63	2+	2.2
S^{32}	2.24	2+	~ 2
A^{40}	1.43	2+	0.9f (1.1f (DWBA))

We find relatively large values of $\beta_2 R_0$ except for the case of argon. (The value for A corrects a numerical slip contained in my earlier paper (*Phys. Rev.* **115**, 928 (1959)).) The low value for A suggests then that a vibrational description is a better way to characterize the collective matrix element in this case. Also shown is the value of $\beta_2 R_0$ (1.1 fermis) that is obtained from a distorted wave Born analysis of Rost and Austern which matched a different experiment with A carried out at Purdue with 18 MeV α-particles.

We also have observed, in this region, odd parity oscillations, i.e. oscillations which are in phase with the elastic angular distribution. When analyzed in this same fashion, as octupole excitations, we obtain the parameters listed in Table 2.

We have expressed the collective matrix elements in terms of $\beta_3 R_0$, but their size suggests that in most of these cases a vibrational description is more appropriate.

TABLE 2. OCTUPOLE NUCLEAR DEFORMATIONS $\beta_3 R_0$ AS OBTAINED FROM ANALYSIS OF α-PARTICLE SCATTERING EXPERIMENTS.

Nucleus	E_x MeV	J^π	$\beta_3 R_0$ fermis
O^{16}	6.13	3−	1.0
Ne^{20}	7.2	3−	1.6
Mg^{24}	6	?	0.5
S^{32}	5	?	∼1
Ca^{40}	∼4	?	0.6

Let me now turn to another region in the periodic table, namely the intermediate nuclei. This is a region of much interest where Cohen found excitation of "anomalous levels" lying typically between 2.5 and 4 MeV above the ground state. As shown in Fig. 6 excitations are indicated in plots of counts versus excitation energy for a given scattering angle. The experiments here shown are those of McDaniels, Chen and Farwell using 41 MeV alpha particles incident on a variety of targets. Peaks according to elastic scattering are prominent. The arrows refer to the well-known quadrupole 2+ levels, strongly excited in Coulomb excitation experiments, and to the so-called anomalous groups. The angular distributions corresponding to these peaks can be measured. What is shown in Fig. 7 is not typical but perhaps our best data, which is for zinc-64. We see the elastic scattering cross-section, the inelastic cross-section going to the 2+ state and the "anomalous" inelastic cross-section which is displaced by a factor of 10 in magnitude; therefore, it is quite comparable to the other curves in magnitude. We observe that the "anomalous" group is in phase with the elastic pattern while the 2+ is out of phase with the elastic. It is then very tempting to ascribe this "anomalous" group to one level per isotope, that being an octupole excitation.

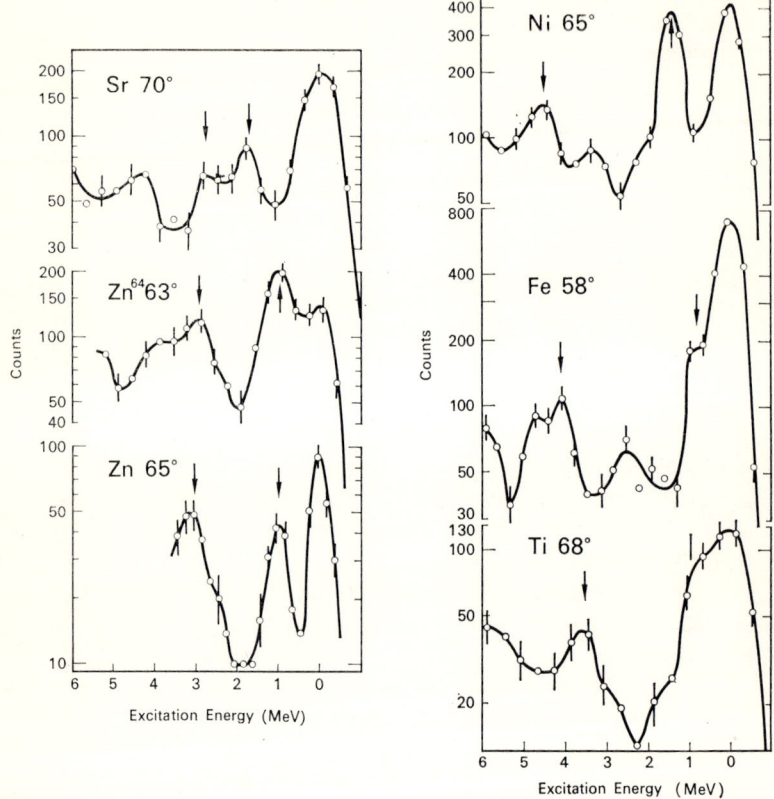

FIG. 6. Alpha-particle spectra showing "anomalous" peaks.

The elements investigated were titanium, iron, nickel, zinc and strontium-88. Strontium-88 was particularly included since excited states which are known to be 2+ and 3− are well separated from higher levels; thus this isotope could be used as a calibration. Although the data in the case of Sr-88 were not as good as those for Zn, Ni and Fe, nonetheless the predicted phase relationships were found. Also I should remark that the 3-assignment has been made

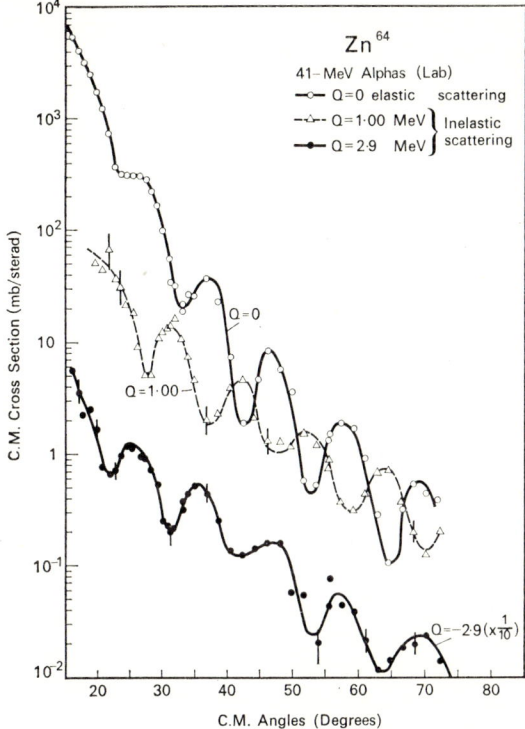

FIG. 7. Elastic and inelastic α-particle scattering from Zn⁶⁴ at 41 MeV.

fairly convincingly by Crut and Wall in some angular correlation experiments for the nickel-58 and nickel-60 isotopes. Similar scattering experiments have been performed by groups at M.I.T., Argonne and Saclay and have been similarly interpreted; I should summarize the data by noting that the predicted phase pattern was found in all cases.

We can again try to estimate the collective matrix elements, the procedure is a little harder in this case because Coulomb complications are quite serious and also because the observed cross-sections fall away from the Fraunhofer, black-edge expression much faster at the higher orders than is the case for lighter target elements. Nonetheless there will be a small range of angle, about 10° in width centered around 25° where the elastic cross-section is close to the Fraunhofer expression. If we normalize in this region, we find values for the matrix elements in the quadrupole case which are close to those obtained from the electromagnetic experiments. Typically the values of $\beta_2 R_0$ are of the order of a fermi. For the octupole vibrations the values of $\beta_3 R_0$ are less, typically of the order of 0.5 fermi. If we couch the matrix elements for octupole excitation in terms of the surface tension C_3 we find that C_3 is in the range 500 to 1000 MeV for the particularly strong excitation in Zn-64 and is usually larger than a 1000 MeV for other elements. The hydrodynamic estimate for C_3 is of the order of 200 MeV. Lane and Pendleberry have given estimates for this parameter of the order of 1000 MeV.

In conclusion I would just like to say that I think that this indicates that α-particle inelastic scattering can be used as a tool to investigate collective excitations; even though the α-particles have $T = 0$ it is hoped that they may reveal other sources of collective excitation. Certainly in inelastic scattering studies with protons such should be the case.

12

Information Obtainable from (p, 2p) Reactions†

I. E. MCCARTHY

I DISCUSS the theoretical interpretation of the $(p, 2p)$ reaction as the direct knockout of a proton from a nucleus by a single-particle interaction with an incident proton. The analysis up to date shows that there is very interesting information about the details of a nuclear reaction available from $(p, 2p)$ reactions at various incident energies, as well as information about the structure of the initial nucleus. It also leaves some interesting questions that have yet to be answered.

The analysis is carried out in the distorted-wave Born approximation. I propose to outline the derivation as completely as I can. However, in order to orient the discussion I would like first to discuss the simplest possible model, the plane-wave Born approximation. In this approximation, the matrix element T_L^M is written in the following way:

$$T_L^M = \int\int d^3r_1\, d^3r_2 \exp(-i\mathbf{k}_L \cdot \mathbf{r}_1)$$
$$\times \exp(-i\mathbf{k}_R \cdot \mathbf{r}_2) v(|\mathbf{r}_1 - \mathbf{r}_2|) \exp(i\mathbf{k}_0 \cdot \mathbf{r}_1) \psi_L^M(\mathbf{r}_2). \quad (1)$$

We are assuming that the knocked-out proton was initially in a single-particle state ψ_L^M. The two-body interaction is approximated by a local, central interaction $v(|\mathbf{r}_1 - \mathbf{r}_2|)$. The kinematic situation is defined in Fig. 1. The k's are the wave numbers corresponding to the E's.

† *Revs. Mod. Phys.* **37**, 388 (1965). This work was supported in part by a grant from the U.S. Atomic Energy Commission and by the University of California Opportunity Fund.

Making the transformation

$$\mathbf{r}_1 - \mathbf{r}_2 = \mathbf{r}$$
$$\mathbf{r}_2 = \mathbf{r}' \qquad (2)$$

we may factorize the matrix element in the following way:

$$T_L^M = \left[\int d^3 r \exp\left[i(\mathbf{k}_0 - \mathbf{k}_L) \cdot \mathbf{r}\right] v(r) \right]$$
$$\times \left[\int d^3 r' \exp\left[i(\mathbf{k}_0 - \mathbf{k}_L - \mathbf{k}_R) \cdot \mathbf{r}'\right] \psi_L^M(\mathbf{r}') \right]. \qquad (3)$$

The first factor is the Fourier transform of the two-body interaction with respect to the momentum transfer suffered by the incident particle. Since we are making the Born approximation, the best choice for $v(\mathbf{r})$ is a pseudopotential that fits free (p, p) scattering, that is a function whose squared Fourier transform is the differential cross-section for (p, p) scattering. This is a momentum space approach. We are using a model for the interaction that reproduces the right momentum components on the two-body energy shell. It is also a model for the interaction off the energy shell. Now, when T_L^M is squared, the first factor is the free (p, p) cross-section.

The second factor is the Fourier transform of the bound-state wave function with respect to the momentum transfer suffered by the residual nucleus or the supposedly inert core. It is the momentum-space wave function of the struck particle.

It is obvious, even though the plane-wave approximation may have only a rudimentary resemblance to the facts, that there are two types of information in the reaction, one about the two-body force, one about the wave function of the struck particle. How do these things affect the angular correlation?

First, the bound-state factor is familiar in plane-wave direct interaction theory. For a surface interaction it is a spherical Bessel function which is zero at zero momentum transfer if $L \geq 1$, maximum at zero momentum transfer if $L = 0$. Specializing to the case $E_L = E_R$, $Q = 0$, $\theta_L = \theta_R = \theta$ for simplification of the discussion, zero momentum transfer occurs when the struck particle was stationary at the moment of impact, that is when $\theta = 45°$.

If θ is varied, we have two spherical Bessel functions in the angular correlation, one for $\theta < 45°$ representing a collision with a particle moving away and one for $\theta > 45°$ representing a collision with a particle moving towards the incident particle. For $L \geq 1$ there are two peaks. Conservation of momentum says they must be of equal height and mirror images about $K = 0$ if plotted against K rather than θ. The width of the spherical Bessel function depends on L and on the rms radius of the wave function.

The first factor influences the angular correlation in the following way. The (p, p) cross-section is greater for a collision with a particle moving away, that is for lower energy in the two-body center-of-mass system. Therefore the left peak is higher than the right.

This approximation thus gives us the idea that the width of the angular correlation distribution is related to the spectroscopic information, the peak-height ratio (for $L \geq 1$) is related to the information about the reaction mechanism.

Let us now consider the derivation of as exact an approximation as possible for the reaction. The transition amplitude for a reaction due to two potentials U and V is

$$T = \langle \Psi^{(-)} | V | \Psi^{(+)} \rangle + \langle \Phi^{(-)} | U | \Psi^{(+)} \rangle, \qquad (4)$$

where $\Psi^{(\pm)}$ are wave functions calculated in the "distorting" potential U. $\Phi^{(-)}$ is the noninteracting wave function for the final state. If the Hamiltonian of the system is the same before the interaction as afterwards, the second term is zero. This is approximately true for $(p, 2p)$ so there is some *a priori* justification for the neglect of the second term. (This is not the case for rearrangement collisions.)

What do we use for $\Psi^{(+)}$ and $\Psi^{(-)}$? $\Psi^{(+)}$ represents an ingoing particle interacting with a nucleus. The distorted wave approximation for this is familiar. It is

$$\Psi^{(+)} = \chi^{(+)} \psi_L^M, \qquad (5)$$

where $x^{(+)}$ is an optical model wave function, $\psi_L{}^M$ is the bound-state wave function discussed before.

$\Psi^{(-)}$ represents a three-body state where two of the bodies have mass 1 and one has mass A. The wave function for this state may be separated into one representing the motion of the center of mass and a two-body wave function which is the solution of the following Schrödinger equation

$$[-(2\mu)^{-1}\nabla_L{}^2 + V_L - (2\mu)^{-1}\nabla_R{}^2 + V_R - A^{-1}\nabla^L \cdot \nabla_R]\Psi^{(-)} = E\Psi^{(-)}. \quad (6)$$

The interaction term $A^{-1}\nabla_L \cdot \nabla_R$ may be treated as a perturbation. In the first order the equation separates into two one-body equations which may be thought of as optical model equations.

$$[-(2\mu)^{-1}\nabla^2 + V]x^{(-)} = E'x^{(-)}.$$

The effective energy is, putting $a = A^{-1}$,

$$E' = \frac{1}{2}\left(\frac{1+a}{1+2a}E_{\text{lab}} + Q\right) \times \left\{1 - \frac{a[(4a^2 + 4a + 2)\cos^2\theta - 1]}{(1+a)^2(1+2a\cos^2\theta)}\right\}. \quad (7)$$

The distorted-wave matrix element is now

$$T_L{}^M = \iint d^3r_1\, d^3r_2\, x^{(-)*}(\mathbf{k}_L, \mathbf{r}_1)x^{(-)*}(\mathbf{k}_R, \mathbf{r}_2)$$
$$\times v(|\mathbf{r}_1 - \mathbf{r}_2|)x^{(+)}(\mathbf{k}_0, \mathbf{r}_1)x_L{}^M(\mathbf{r}_2). \quad (8)$$

I describe a calculation by Lim and myself[1] starting from this matrix element. Unless I specifically say otherwise, I refer to the coplanar symmetric case. We attempted to use models for every factor in the matrix element which were as realistic as possible and which were directly related to simpler experiments so that there are no free parameters in the theory.

The optical model potentials were found by interpolation in the available optical model literature. Spin–orbit coupling was omitted, since we are not describing polarizations. This approximation is good for momentum transfers which are not too large. The reactions we are describing have in fact quite small momentum transfer. (I discuss this later.) The parameters for the entrance and exit channels, respectively, were V_0, W_0, V_1, W_1 with an Eckart form factor described by r_0 and b.

The bound-state wave function description is most important. The information we have, assuming a single-particle model, is the binding energy from the $(p, 2p)$ experiment itself, and the rms radius of the charge distribution. This gives us a strong set of constraints since for light nuclei we have both the s- and p-state angular correlations so that we must fit the electron-scattering radii.

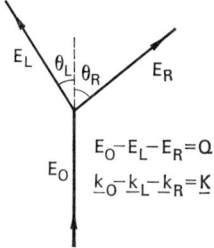

FIG. 1. Definition of the symbols used in the description of the $(p, 2p)$ reaction.

In fact calculations with potential wells of different shapes have shown that with a given binding energy and rms radius the matrix element is essentially independent of the well shape, provided it has a fairly flat bottom. We used a square well for simplicity with parameters V_B and a.

For the two-body pseudopotential we used

$$v(r) = -83 \left(\frac{\exp(-0.73r)}{0.73r} - 5 \frac{\exp(-1.5r)}{1.5r} + 20 \frac{\exp(-3r)}{3r} \right) \text{MeV}. \tag{9}$$

The squared Fourier transform of this $v(r)$ gives the fit to the 90° differential cross-section for (p, p) scattering shown in Fig. 2.

Figure 3 shows the result of the calculation for 155-MeV incident protons on C^{12} compared with experiment.[2]

There are two things to be noticed about the result. First, the magnitude of the left peak is fairly well reproduced. Less exact single-particle direct interaction theories have not previously reproduced magnitudes anywhere near correctly and this is a very hopeful indication that we are on the right track. Second, the size of the right peak is underestimated by a factor of about 2. Our

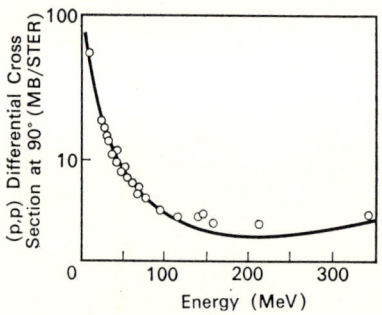

FIG. 2. The differential cross-section for (p, p) scattering at 90° calculated from the 3-Yukawa pseudopotential[10] plotted against experimental data.

intuition from the plane-wave model would tell us that we have not enough high momentum transfer components in the two-body interaction.

We have varied the factors in the integrand of the matrix element to see if the correct angular correlation can be produced without using hopelessly unrealistic values of the parameters. Our experience with zero-range calculations[3] has shown that, although distortion does reduce the right peak relative to the left peak, the large observed effect certainly cannot be reproduced with any believable optical model parameters.

The binding energy and rms radius of the bound-state wave

function are fixed, but we have artificially added a 20 per cent admixture of a $2p$ wave function in an infinite square well of the same radius in order to see if additional high momentum components in the bound state will help. This admixture changed the differential cross section by about 1 per cent.

However, as expected, the ratio of peak heights turned out to be very sensitive to the amount of one pion component in the interaction. Reducing the ratio of one pion component by a little less

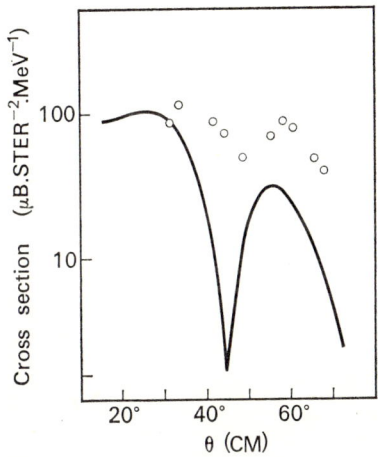

Fig. 3. 155 MeV $(p, 2p)$ angular correlation for the p state of C^{12} compared with theory.

than a factor of 2 produces a reasonable fit to the experiment. The fit for a factor of 2 is shown in Fig. 4.

It might be asked whether the spin dependence of the two-body force might not change the result appreciably. We have carried out an explicitly spin-dependent formulation of the problem. In the cross-section there are terms which contain space-symmetric and space-antisymmetric matrix elements.

In the zero-range DWBA, the space-antisymmetric terms vanish for the case of symmetric coplanar scattering. With our pseudo-potential the antisymmetric cross-section calculated without spin–orbit coupling is 1 per cent of the symmetric cross-section at 155 MeV and 20 per cent at 50 MeV. We can neglect it at 155 MeV. The size of the antisymmetric cross-section is a measure of how far off the energy shell we are. Spin–orbit coupling is small at 155 MeV.

A preliminary formulation of the problem has been made with a velocity-dependent two-body potential. In this case, the separation approximation is not equivalent to the impulse approximation

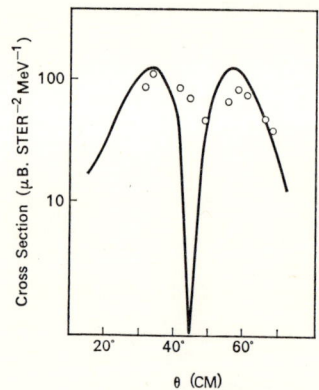

FIG. 4. The angular correlation for 155 MeV $(p, 2p)$ on the p-state of C^{12} with the core terms in the interaction increased by a factor of 2.

and it is possible for the angular correlation to be different at 155 MeV from the impulse approximation value. It is shown later that the impulse approximation is quite close for the local pseudo-potential at this energy. A calculation using simplified optical model wave functions will be performed with spin–orbit coupling and velocity dependence. Of course it must be realized that the local pseudopotential is chosen in such a way as to simulate the velocity dependence in free (p, p) scattering.

The questions we are left with are: Is it necessary to use a better approximation than the distorted-wave Born approximation to describe the reaction? If a better approximation does not give a significantly different result, can the difference between the two-body interactions inside and outside nuclear matter be explained by a theory of nuclear matter? These questions are left for the future, but it is possible to make one qualitative observation.

Momentum transfers under discussion here are not much greater than $1F^{-1}$. This is the reciprocal of the healing distance at which, at least for infinite nuclear matter, the two-body wave function becomes like the wave function with strong short-range correlations. This at least indicates that the strong correlations are not likely to upset a simple two-body theory like the distorted-wave Born approximation for the momentum transfers we are concerned with.

The present theory is sufficiently complete to answer some more questions of principle. It may be asked why it is useful to use a pseudopotential which is a model for the two-body interaction both on and off the energy shell, when we only know the two-body interaction on the energy shell. Is not the impulse approximation adequate? We can answer this question by comparing our result with the impulse approximation for the same situation.

We first make the separation approximation. That is we use eq. (3) with the second factor replaced by the zero-range distorted-wave matrix element. The first factor is replaced by the appropriate expression on the two-body energy shell. Figure 5 shows how the impulse approximation compares with the more exact theory at 155-MeV and 50-MeV incident energy.

Although the impulse approximation is not bad at 155 MeV, it is very far off at 50 MeV. This leads to the useful conclusion that $(p, 2p)$ experiments at 50 MeV contain a large amount of information about the effective two-body force in nuclear matter off the energy shell. $(p, 2p)$ experiments are a very good way of getting this information because they are particularly simple experiments with a three-body final state. The fact that one of the bodies is heavy enables us to use the separable perturbation expansion of eq. (5).

So far I have not said much about the information that can be obtained about bound states. Our early zero-range calculation[3] showed that the rms radii of the s and p states in light nuclei can be well enough determined to show that, if the square-well parametrization of the wave functions is used, the s state well is narrower and deeper than the p state well. This state dependence of the single-particle potential also arises from the finite nucleus calculations of Brueckner, Lockett, and Rotenberg.[4] The finite range calculation bears this out. The s-state rms radius appears to be a little larger than that of the α particle.

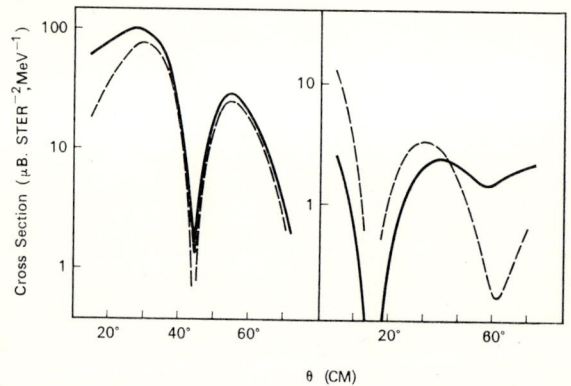

FIG. 5. Comparison of the impulse approximation (broken line) with the more exact theory at 155 MeV (left) and 50 MeV (right).

The calculation also contains information about the adequacy of simpler methods of analysis. The 155-MeV results show that at higher energies the impulse approximation should be quite good enough for obtaining spectroscopic information. Jackson and Berggren[5] have compared a simple zero-range distorted wave calculation at 170 MeV for a very light nucleus, Li⁶, with a plane wave calculation which is modified by introducing space weighting factors related to the WKB approximation. The agreement is good enough to show that at very high energies, say greater than 200 MeV, a modified plane wave approximation in conjunction with the impulse approximation is sufficient to tell us about the bound-state properties.

Finally, I would like to mention one experimental point. It is noticeable that the low momentum transfer minimum in the angular correlation is much deeper in theory than in experiment at 155 MeV (although at 50 MeV the distortion turns the minimum

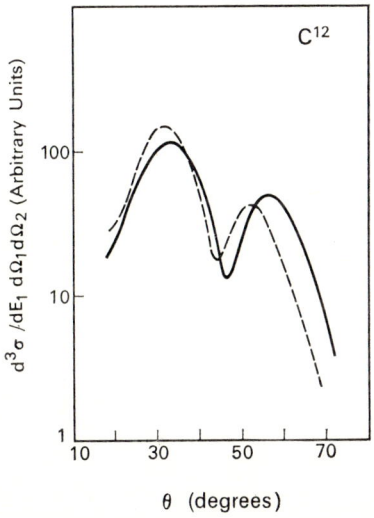

FIG. 6. Comparison of the angular correlation for 155-MeV protons on the p state of C^{12} for $E_L = E_R = 69.5$ MeV (full line) with that for $E_L = 74.5$ MeV, $E_R = 64.5$ MeV (broken line), showing how poor energy resolution can fill in the minimum.

into a maximum just as it does for 2+ excitations by proton inelastic scattering at low energies). Good angular resolution does not change this. However, if only the sum of the final energies is well resolved, there is an explanation. Unequal final energies with the same sum can shift the minimum considerably as is shown in Fig. 6. Here the curve for $E_L = E_R$ in the zero-range approximation is compared with that for $E_L = 74.5$ MeV, $E_R = 64.5$ MeV.

We conclude from this that individual resolutions of at least 1 MeV are needed.

Notes

[1] K. L. Lim and I. E. McCarthy, *Phys. Rev. Letters* **13**, 446 (1964).

[2] J. P. Garron, J. C. Jacmart, M. Riou, C. Ruhla, J. Teillac, and K. Strauch, *Nucl. Phys.* **37**, 126 (1962).

[3] K. L. Lim and I. E. McCarthy, *Phys.* **133**, B1006 (1964).

[4] K. A. Brueckner, A. M. Lockett, and M. Rotenberg, *Phys. Rev.* **121**, 255 1961).

[5] D. F. Jackson and T. Berggren (preprint).

Index

Accelerators 6, 48, 58
Alpha particle
  early (α, p) reactions 5, 144
  elastic scattering as nuclear probe 3, 127
  nuclear size from α-decay 131
  reactions at high energy 272
  scattering from collective states 106, 298

Collective models, relation to shell model 87, 91
Compound nucleus model 18, 50, 52, 153, 229
  assumption of independence of formation and decay 18, 33, 153
  evaporation model 38
  time-dependence 42

Deformed nuclei
  Hartree–Fock calculations 87
  inelastic scattering 108, 302
  parameters determined by inelastic scattering 302
  rotation model 89
Deuteron
  optical model parameters 67
  reactions at high energy 272
  scattering from collective states 304
  stripping *see* Stripping
Diagrams
  definition for elastic scattering 76
  explanation of Born approximation for stripping 94
Direct reactions 50, 92

Distorted wave approximation
  inelastic scattering 102, 106, 109, 111, 307
  $(p, 2p)$ reaction 117, 318
  resonances 15, 144, 158
  stripping 98
Distorted waves
  approximate wave function 73, 105
  magnitude and phase 69
  parameters for stripping 99

Elastic scattering
  approximate wave function 73
  average cross-section 41, 53, 233
  complex potential and mean free path 49, 217
  compound elastic scattering 18, 40, 54, 231, 236
  continuum model 41, 44, 50, 226
  diffraction scattering 46, 223, 298
  Hauser–Feshbach theory 40, 64, 245
  phenomenological optical model fits 60–68
  polarization 61
  potential scattering model (early) 10, 137
  resonances, giant (shape) 15, 54, 137, 224, 243, 254
  Rutherford scattering as nuclear probe 3, 8, 127
  t-matrix 75
  theory of potential scattering 11, 138, 241
  wave functions and particle flux 69

INDEX

Electron scattering 9, 82
Experiments, general discussions 46, 58

Fluctuations in cross-sections 36

Independent particle model 51
  approximate real potential derived from properties of the nucleon–nucleon interaction 82
  explanation of approximate validity for nuclear matter 81
Inelastic scattering
  adiabatic approximation 299
  collective mechanism 103, 300
  coupled channels calculation 110
  diffraction scattering model 104, 301
  distorted wave Born approximation 102, 106, 109, 307
  distorted wave t-matrix approximation 111
  generalized optical model 108, 111, 299
  identification of direct mechanism 100
  impulse approximation 111
  nucleon-collision mechanism (microscopic description) 103, 107, 113
  phase rule 302
  plane wave Born approximation 101, 298
  rotational mechanism 108, 302
  shapes of angular distributions 102, 104, 105, 109, 302
  vibrational mechanism 109, 303

Liquid drop model 51

Mass relationships of nuclei 4, 126, 132
Mean free path and complex potential 49, 217, 273

Neutron
  anomalies in low-energy scattering and capture 17, 152, 165, 182
  elastic scattering 44, 48, 54, 66, 153, 216
  significance of discovery 6
  total cross-section 222, 224, 229, 275
Nuclear matter
  Brueckner–Bethe–Goldstone equation 80
  explanation of approximate validity of independent particle model 81
  Fermi gas model 82
  k-matrix of Brueckner 80
  real potential depth 82
  semi-empirical mass formula 79
Nucleon–nucleon interaction 77
  pseudopotential for 90° p–p scattering 317
  relation to imaginary nucleon–nucleus potential 49, 217
  relation to real nucleon–nucleus potential 82, 217
Nucleon–nucleus potential
  approximate derivation of imaginary part 49, 217
  approximate derivation of real part 82, 217

Optical model
  energy-dependence 65, 69
  generalized for collective excitations 108, 111
  integral equation and diagrams 74–76
  non-local 68
  phenomenological fits to entrance channel data 60–68
  potentials related to nucleon–nucleon scattering 49, 82, 217
  reaction cross-section 64, 218, 231, 234
  relation to resonances 53, 231
  strength function 56, 111
  transmission coefficients 40, 64
  WKB approximation 220
  wave functions and particle flux 69–74

$(p, 2p)$ reaction 113, 313
  angular correlation 115, 318
  distorted wave t-matrix approximation 117
  for identifying collision mechanism 103, 113
  impulse approximation 115, 314, 321
  probe for nucleon momentum distribution 113
  single-particle eigenvalues 114
  single-particle r.m.s. radii 116, 322
  spectroscopic tool 117
  widths of single hole states 114
Polarization in elastic scattering 61
Proton
  elastic scattering 61, 64
  inelastic scattering 107
  optical model parameters 65

Resonances
  averaging 231, 233
  Breit–Wigner theory 21, 166
  distribution of reduced width amplitudes 35, 192
  distribution of spacings 35
  for thermal neutrons 159
  in (α, p) reactions 5, 137, 144
  relationships between widths and spacings 55, 237, 247, 259
  shape resonances in elastic scattering 15, 54, 137, 224, 243, 254
  strength function 56, 111, 259
  theory of resonance expansion and definitions of parameters 24, 168, 238
  two-channel distorted-wave theory 15, 144, 158
Rotational model 89

Shell model
  Hartree–Fock calculations 87
  nuclear forces 86
  relation to collective models 87, 91
  single-particle model 83, 229
  state dependent Hartree–Fock potentials 117
Single-particle models 52, 53, 83
  experimental determination of eigenvalues 98, 114
  experimental determination of r.m.s. radii 116
Spectroscopy 5
Strength function 56, 111
Stripping 51, 93, 276
  angular momentum determination 96, 288
  Born approximation 94, 278
  Butler's theory 95, 106, 277, 286
  diagrams for Born approximation 94
  diagrams for DWBA 98
  distorted wave Born approximation 98
  single-particle eigenvalues 98
  spectroscopic factors 97

t-matrix, definition for elastic scattering 75
Time-dependence in scattering 42–44
Transmission coefficients 40, 64

Unified model, adiabatic approximation 88

Vibrational model 88
  for inelastic scattering 109